GW00722508

JEAN-GABRIEL JEUDY • MARC TARARINE

THE JEEP

"The Jeep, the Dakota and the landing craft were the three tools that won the war."

General Eisenhower
Supreme Commander of the Allied Forces in Europe

INTRODUCTION

There is something irresistible about an idea whose time has arrived. The idea for the Jeep is almost as old as the motor vehicle, but it was not until 1940 that everything came together and quite suddenly the nimble little car was launched on a career which had a not inconsiderable influence on the course of World War II.

The authors of this book give us a striking example of a precursor of the Jeep in France as early as 1906. During World War I the Model T Ford successfully performed the kind of task at which the Jeep excelled 30 years later. During the thirties another classic design, the Austin Seven, was used by both British and German armies and tested by the American forces for the duties the Jeep was later to perform.

There were some exotic excursions up blind alleys with devices like the Motor Cart and the Belly Flopper and in some quarters enthusiasm for the motor cycle and sidecar remained strong, but the idea for the Jeep was gathering momentum and the American Army laid down the basic specification. The achievement of Karl Probst and his little team at Bantam in designing and constructing the first prototype in two months in the summer of 1940 is still awe inspiring. But that was not all. By the 75th day from the start, 70 prototypes had been delivered. This was the Jeep. Modifications were made as a result of experience, but the basic concept remained unchanged. Yet sadly those who gave it form and substance were denied the due reward for their efforts.

The design was simple. Deceptively so, for when the war was over, a number of people thought they could improve upon it, and vast sums of money were dissipated in the attempt. Vehicles like the Delahaye VLR and the Austin Champ incorporated the latest features of automobile design but they failed miserably and expensively. After years of frustration, the French Army put the Jeep into production once more. Many of the great names in the industry have been associated with unsuccessful efforts to replace the Jeep. The Land Rover has been the most successful because Maurice Wilks did not try to prove that the original concept of the Jeep was wrong; he simply addressed himself to avoiding its main drawbacks, an excessive thirst and a tendency to rust.

Forty years on, the basic Jeep formula is still valid.

JEAN-GABRIEL JEUDY • MARC TARARINE

THE JEEP

Translated by Gordon Wilkins

FREDERICK WARNE

Printed in France

First published in France under the title :
La Jeep Un défi au temps
Layout : François Abegg
ISBN 0 7232 2872 8

CONTENTS

ANCESTORS OF THE JEE

When the first horseless carriages made their appearance a few far-sighted individuals immediately saw what an advantage they could be to the army.

In all the countries smitten by the new passion for motoring, committees were set up to interest their armed forces in using petrol driven vehicles. It was in Europe, particularly in France and Germany that the movement gained its earliest successes. The first big maneuvers in which motor vehicles took part were those in the French South West in the autumn of 1897. On this occasion, four de Dion Bouton tricycles and three Panhard Quadricycles with Daimler engines were placed at the disposal of the VIth and XVIIIth corps. However, these vehicles did not actually belong to the Army. They were civilian vehicles whose owners, all volunteers, had been mobilised with their machines for the duration of the maneuvers. This custom and the habit of hiring vehicles from various manufacturers for the great maneuvers continued in France, England and Germany right up to the first World War.

It was in 1899, that the French Army bought its first three cars, a Panhard for the Vincennes depot, a Maison Parisienne for Versailles and a Peugeot for Besançon.

From the beginning, the army was chiefly interested in using the cars for reconnaissance and liaison. As an example there were the trials undertaken in October 1897 by the technical service of the Artillery which provided for a run of 200 km out and back starting from the Place St Thomas D'Aquin in Paris, headquarters of the Artillery Board.

Thus the military were not interested in the automobile as a means of urban transport, which it was at that time for most car owners, but rather as a vehicle to run long distances on the country tracks (one could not call them roads at that time), to permit officers to carry out reconnaissance or to establish liaison between army corps. Moreover, once they had got over their surprise, the military did not hesitate to use the automobile in all parts of the world. General Gallieni, for example was provided with a motor vehicle in Madagascar. In 1904, Captain Gentil of the Vincennes workshops converted a Panhard for armed reconnaissance. This vehicle, which was armed with a Hotchkiss machine gun, had a good performance over

The motor machine gun devised by Captain Gentil seen here at the wheel in the yard of the Panhard factory in the Rue d'Ivry in Paris in 1906. It embodies the fundamental tactical ideas which were foremost in the concept of the Jeep.

varied terrain thanks to the power of its engine and the flexibility of its wooden chassis. It was used successfully in Morocco to reconnoiter routes in hostile territory, for journeys by General Lyautey and for carrying despatches.

It should really be regarded as the very first reconnaissance vehicle suitable for operational use, with its tourer body, a machine gun, a crew of three, able to go anywhere and capable at the same time of keeping up a good pace on the road. It required little maintenance and being used to obtain information, carry orders and take officers to the scene of the action, it was more than just the idea for a Jeep, it was already a Jeep.

In 1910 Davidson fitted out two Cadillac chassis for the same purpose in the United States. On the same theme, the long letter sent by Major Thomas J. Dickson to Hupp Motors in 1912 on the subject of the Hupmobile in service at Fort Rilley in Kansas (published by Bart H. Vanderveen in Old Motor for May 1966) says much the same thing. The Major was proud of the performance of the car which allowed officers to follow the course of the maneuvers whatever the country, even where *"the mules, which according to military tradition are the personification of stubbornness"* had come to a standstill.

The uses of the car were those of a Jeep and one finds the same requirement expressed time after time. The vehicle should be able to go anywhere where the mission required men to go, regardless of the terrain and the conditions.

The idea of the Jeep is thus as old as the use of light vehicles by the military. This is one of the reasons why the history of its creation is so complicated and why in spite of all the claims made in this connection no one invented the Jeep.

As we know, the first World War made great demands on car manufacturers throughout the world. Ford produced tens of thousands of his Model T for the Allies. In the Middle East and especially in Mesopotamia, the Ford Model T was used as a patrol and reconnaissance vehicle armed with a Lewis machine gun and fitted with a crude wooden body. The *"Spider"* had a good performance over difficult country and carried out its job extremely well, patrolling many routes in the desert. In the USA it served

This Model T Ford fitted with a Vickers machine gun was used by the British in Mesopotamia and Egypt for reconnaissance and desert patrol. It is more than the idea; it is already a Jeep 25 years before the real one arrived.

as the standard of comparison for all designs for light vehicles for many years after the war.

Once peace returned, the victors no longer talked about designing new equipment. On the one hand the funds available had to be used to reconstruct the economic infrastructure and on the other hand such quantities of equipment had been produced during the hostilities that there was an excess of stocks rather than a shortage.

It was only at the end of the twenties that armies which had been using their war-time equipment up to then, began placing significant orders for new vehicles. For the most part they contented themselves with adapting civilian vehicles already in production.

At this time the fashion was for ultra light vehicles which, with most of all their bodywork removed, had an adequate performance cross country. Among them was the Baby Austin or Austin Seven which had a busy military career. It was used by the British Army; it was also built under licence in Japan and by BMW in Germany where it equipped the Reichswehr which used it to train the personnel for its future motorised units.

In the United States, a Seven built at Butler by

In 1928 Berliet designed a four-wheel drive liaison vehicle. The prototype, rather different from the two vehicles later built for the French Army, had a semi-forward driving position. The six-cylinder 40 b.h.p. petrol engine was mounted at the rear. Suspension was by four cantilever springs.

For reconnaissance the Reichswehr used a lot of BMW Dixis which were none other than Austin Sevens built under license in Germany. Their low weight gave them a creditable cross country performance.

The Adler Favorit and Wanderer W II on the test tracks of the Doeberitz military establishment near Berlin where the Reichswehr trained thousands of cross-country drivers (November 1932).

American Austin was fitted with a triple machine gun mounting.

Later on, its engine was also used to drive the Belly Flopper. We shall return to the Butler factory all through the history of the Jeep because in 1935 the American Austin became the Bantam.

In spite of their good points these little cars were not very suitable for use on rocky or very uneven ground, so manufacturers began developing vehicles specifically for the purpose which were relatively complicated mechanically and therefore expensive. Berliet, Laffly, FN, Vidal und Sohn and many others must be given credit for venturing into this field but one cannot say that they were encouraged by the orders they received.

With a few exceptions, no specific all-terrain liaison vehicles were bought at that time, either because they cost too much in a budgetary period which was not favorable to military expenditure or because their maintenance seemed too difficult for men from the ranks who were not specialists. Furthermore, the view at that time was that partly because of the absence of big mechanised units, these vehicles could only cover a small part of the requirements. Most missions could be accomplished by normal cars which had been requisitioned. In England several companies specialising in the hiring of military equipment were set up, like the Artillery Transport Company or Mechaniza-

An American Baby Austin with triple machine guns and balloon tyres. The steering wheel could be removed to help in stowing the mounting. (1933).

The British Army used the sturdy little Austin Seven and also the Morris Minor seen here at Saarbrucken in 1935 with radio equipment.

Based on the Fiat Balilla, the "Spider 508 Militare" was built for the Italian Army from 1932 to 1937. Unlike the civilian model, the 508 M had a four-speed gearbox and a lower axle ratio.

Seen in front of the US Embassy, Rio de Janeiro, this light Chevrolet Car was used by the U.S. Marine Corps in small numbers in 1929-30. A water-cooled Browning 30 Machine gun could be mounted on brackets either in front or rear.

tion Ltd. to provide equipment for the Army, especially during big maneuvers.

Thus Italy chose a militarised version of the popular Fiat Balilla Spider to equip its army from 1932 onwards, holding that it was the most popular vehicle in Italy and therefore the one which could be obtained in the greatest numbers by requisitioning. Things continued very much in this way until the middle of the thirties when projects for the motorisation of large units were put into practice. The European military were still more interested in achieving cross-country transport for anti-tank or anti-aircraft weapons, or for crews using heavy automatic weapons, than in making available to the great mass of troops a light vehicle capable of overcoming all obstacles and undertaking a great variety of missions. So what interested them was a specialised type of vehicle closer to the Dodge than the Jeep.

The Tempo G 1200 with two 19 b.h.p. ILO two-cylinder engines, one in front, one at the rear, was an original design by Otto Dans for the German company Vidal und Sohn. Although it was never adopted by the German Army, this vehicle was a considerable success in its day and from 1936 on was exported to about forty countries.

The Mercedes-Benz G 5 152 with four-wheel drive and four-wheel steering, 320 of which were built between 1937 and 1941. It is typical of the cross-country liaison vehicles used by the German Army before the arrival of the Volkswagen Type 82 Kubelwagen.

Apparently inspired by the Tempo, the Type 47 light tractor built by the Fabrique Nationale at Herstal in Belgium never got beyond the prototype stage. It had one engine in front and another at the rear.

The single DAF MC 139 amphibian built in 1939 was a very original design with an engine from a front-wheel drive Citroen mounted transversely between the two driving positions. It had four-wheel drive and all wheels could be made steerable.

The 508 C Militare based on the Fiat Balilla 1100 was produced in a colonial version with civilian bodywork, then from 1939 with a military tourer body. This was the standard liaison vehicle of the Italian Army right through the Second World War. Driven by a four-cylinder engine giving 30 b.h.p. at 4,400 r.p.m. it had only rear-wheel drive.

The Simca 5, French version of the Fiat Topolino, was never intended for military use but considerable numbers were requisitioned in 1939. Tank Corps officers particularly appreciated their compact dimensions which helped them to insinuate themselves into small spaces, while the folding top allowed the officers to stand up and supervise the operations of their armoured vehicles. They were also appreciated by the German Army, witness this one captured by the British at Bayeux on June 8 1944.

Military opinion, at least in Europe, was not yet ready to accept a general purpose cross-country vehicle. On the contrary, machines were put into service which had been designed for a specific purpose, with bodywork, gear ratios and internal arrangements which for the most part prevented their effective use for any purpose other than that for which they had originally been conceived. We were still far from total motorisation and the concept of a general purpose vehicle.

During the first years of the second World War it was the motor cycle, especially one with sidecar wheel drive, and not a real four-wheeled vehicle, which was entrusted with reconnaissance missions and general liaison.

Although they were dangerous for their crews and only able to cross muddy or uneven ground with difficulty, the motor cycles and sidecars rendered good service before they were replaced, usually by the Jeep or the Volkswagen Kubelwagen.

As was claimed in a notice on the Laffly stand at the first Paris post-war motor show, the V 10 M should have become the French Jeep. A production order was issued but could not be carried out because of the events of May 1940. The four wheels drove and steered.

4,800 units of the Kurogane Type 95 were built for the Japanese Imperial Army. It was one of the rare Japanese 4 × 4 vehicles and one of the few of entirely Japanese design not inspired by American vehicles. Its two-cylinder engine only developed 25 b.h.p. but it had a low fuel consumption of about 8 lit/100 km.

A Volkswagen Kubelwagen of the Afrika Korps in Libya fitted with sand tyres.

"America's greatest contribution to modern warfare"
General George C. Marshall

THE HISTORY OF THE JEEP

The origins, paternity and production of the Jeep have given rise to many arguments ans a great deal of controversy which culminated in a lawsuit in 1948. To find its real origins we must go back to the first World War, where experience on the muddy pock-marked battlefields of France brought home to the military the potential advantages of a small versatile motor vehicle capable of carrying a machine gun with its crew and ammunition. This simple idea was the underlying theme in all the experiments and design studies which led to the Jeep.

Who invented the Jeep and when? A certain unanimity seems to exist in military circles to the effect that no one really invented it. At the end of a far-ranging analysis by all the Americans involved, the Jeep emerged as the logical, indeed the irresistible outcome of a process of evolution guided by experience, and orientated towards the needs of the fighting forces. It was given its final form by the methods traditionally used in the US Army to define the characteristics of the vehicles it intends to order.

This theme has been widely developed in articles, memoirs and studies on the subject by high ranking officers.

"The Jeep was a universal idea which no one person invented, created or developed"
General J.W. Curtiss

"The Jeep is an evolution and not an invention."
General J.S. Barzynski

"The Jeep is the fruit of specifications defined by the military over a long period."
General R.M. Danford

Colonel Van Deusen, who ran the office dealing with transport vehicles at the US Ordnance summed it up by saying "The idea of the Jeep originated with the infantry who needed a low, very powerful vehicle with four-wheel drive."

Colonel Robert G. Howie, who was later to be involved in the development of the Jeep, recalled the problems involved in moving heavy automatic weapons, which had to be dismantled each time they changed position to lighten the burdens on individual infantrymen. Howie, who had been through it all in the first World War, realised that a compact vehicle with a low silhouette, capable of carrying a heavy machine gun and two men over any kind of terrain would be an invaluable asset. The whole idea of the Jeep was there. Such a vehicle would have to be capable of replacing the mule, the horse or even the motor cycle and would need to be designed specially for these requirements.

In 1919 the Technical Division of the US Ordnance and the Technical Committee of the Quarter Master Corps recommended the acquisition of a reconnaissance vehicle designed to travel over all types of terrain, with a low profile to reduce its vulnerability and sufficient ground clearance for use over rough country or damaged roads. It should have a low unladen weight, with an adequate carrying capacity and should be capable of fording streams or rivers. During the twenties the American Army tested a variety of vehicles built according to this general concept.

In 1921 fifteen prototypes of light vehicles fitted

This Ford Model T was tested in 1923 by the Ordnance Department which was then responsible for trials of US Army vehicles except fighting vehicles. The tests of this "reconnaissance tractor" were considered successful at the time and figure among the most significant leading up to a future all-terrain light reconnaissance vehicle. The Ford had lost its bodywork but gained balloon tyres and a drive to the front wheels.

with Caterpillar tracks were tested at the Aberdeen Proving Ground to assess their capabilities. They were particularly intended for reconnaissance in place of motor cycles and sidecars which were not thought to have sufficient performance over difficult country.

Trials on these fifteen vehicles at Aberdeen showed that they were much too heavy and as a result the design studies were directed more towards light commercial vehicles. Developments in this new line of research were largely due to the ingenuity of William F. Beasley, an engineer and Capt. Carl Terry.

In 1923 they turned their attention to a Ford Model T. They stripped it of everything which was not essential and got its weight down to 1,100 lb (500 kg). It performed better than the heavier machines, but the standard tyres which had been retained performed badly in mud or sand. Capt. Terry therefore suggested that they should fit it with tyres taken from surplus aeroplanes stored at the Phillips airfield.

In his memoirs W.F. Beasley recalls "*We cut the wheels to adapt them for small-size airplane tires. That is how the idea originated for what were later to be known as balloon tires. This modification enormously improved the traction. We fitted a body with two seats protected from the wind by a simple tilt. The machine which resulted was the lightest thing of its kind ever devised up to that time. It contained within it the basic concept of what was later to become the Jeep.*"

Unfortunately the project got nowhere because in trying to satisfy all the potential users, who each

The Motor Cart, built in 1923, at Fort Benning. The driver walked behind the machine. The V - twin engine came from a Harley-Davidson motor cycle. This approach to the light all-terrain vehicle had no immediate result but the Mechanical Mule produced in quantities from 1956 for the Marines was a direct descendent.

wanted to add their own special equipment, it put on too much weight, which ruined the whole idea.

At the beginning of the thirties, Arthur W. Herrington, Co-President of the Marmon Herrington Company of Indianapolis became convinced that the Army, which was then thinking about becoming more mechanised would have a need for light, very mobile cross country vehicles.

With this in mind, his company, which had already acquired a good reputation for converting trucks into cross-country vehicles, fitted a Ford 1 1/2 tonner with four-wheel drive in 1934. Herrington interested the Belgians in the idea and supplied them with a converted Ford in 1936 but it turned out to be too heavy. Herrington therefore turned to the half-ton Ford, starting in July 1936, and in September trials were organised at the King Ranch in Texas.

On June 26, 1937, the Defence Department acquired five Marmon-Herrington-Ford 4 × 4 half tonners (Model LD I) which were tested intensively at Fort Benning early in 1938. With a low silhouette, a 990 lb (450 kg) payload and a speed of over 35 m.p.h. they were better than anything the Army had been able to obtain up to then and on January 20, 1939 an order for 64 was issued via the QMC depot at Holabird.

This machine, which can also be regarded as an ancestor of the Jeep, was nicknamed *"Our Darling"* and later *"The Grand-daddy of the Jeep."*

During the same period (1934) Robert G. Howie became interested in the idea of developing a powerful vehicle with a low profile.

Later, in January 1937, working with Sgt. Helvin G. Wiley he devised a very low vehicle for use over rough country. The machine which was completed in March, was equipped with an Austin Seven four-cylinder engine supplied by Bantam. It was only 33.25 in high and Howie described it as a "snake in the grass". The engine was mounted at the rear and drove the front wheels. It could carry two men in a prone position, but the absence of shock absorbers made it tiring to drive and earned it the nickname of the *"Belly Flopper"*. It weighed 460 kg, had a wheelbase of 1.90 m and a top speed of 45 km/h. Armament consisted of an 0.30 (7.62 mm) machine gun with 1,500 rounds of ammunition.

The Army tested the *"Belly Flopper"* in April 1937 but enthusiasm was somewhat muted. The low build certainly helped to make it inconspicuous on the battlefield, but appreciably restricted its ability to travel across country. Major John C. Dotson, an instructor in the automobile section at Fort Benning remarked *"The only way to get it to where it was needed was to put it on a truck."* In fact the contraption proved unsatisfactory and was abandoned.

In March 1940, at the invitation of General Walter C. Short, Delmar Ross, Chief Engineer of Willys and Joseph W. Frazer, the President of the company, were present at a demonstration of the *"Belly Flopper"*. Ross immediately saw the possibilities of such a vehicle and realised that this one could be regarded as a step in the evolution of something with a better performance.

In 1938 the Bantam company lent three standard Austin roadsters to the National Guard in Pennsylvania. The object of the loan was to demonstrate the capabilities of the cars and the possibility of adapting them for military reconnaissance. The National Guard was able to try them during its summer maneuvers and reported favourably on them. Charles Payne, who was in charge of Bantam sales to the Army, took the opportunity to interest the American authorities in the idea of ordering a special reconnaissance vehicle based essentially on these roadsters.

The Belly-Flopper built by Master Sergeant Melvyn C. Wiley to the ideas of Colonel Robert G. Howie. With this vehicle the real development of the Jeep began. The Belly Flopper, built with salvaged parts, had a Baby Austin engine from a Bantam and was shown to General W.C. Short and Mr. Delmas Roos of Willys.

This drawing by Lt. H.G. Hamilton appeared in the American Cavalry Journal in 1935. The article, entitled "A light cross country car", it often regarded as the first clear expression of the military need for a vehicle of this type.

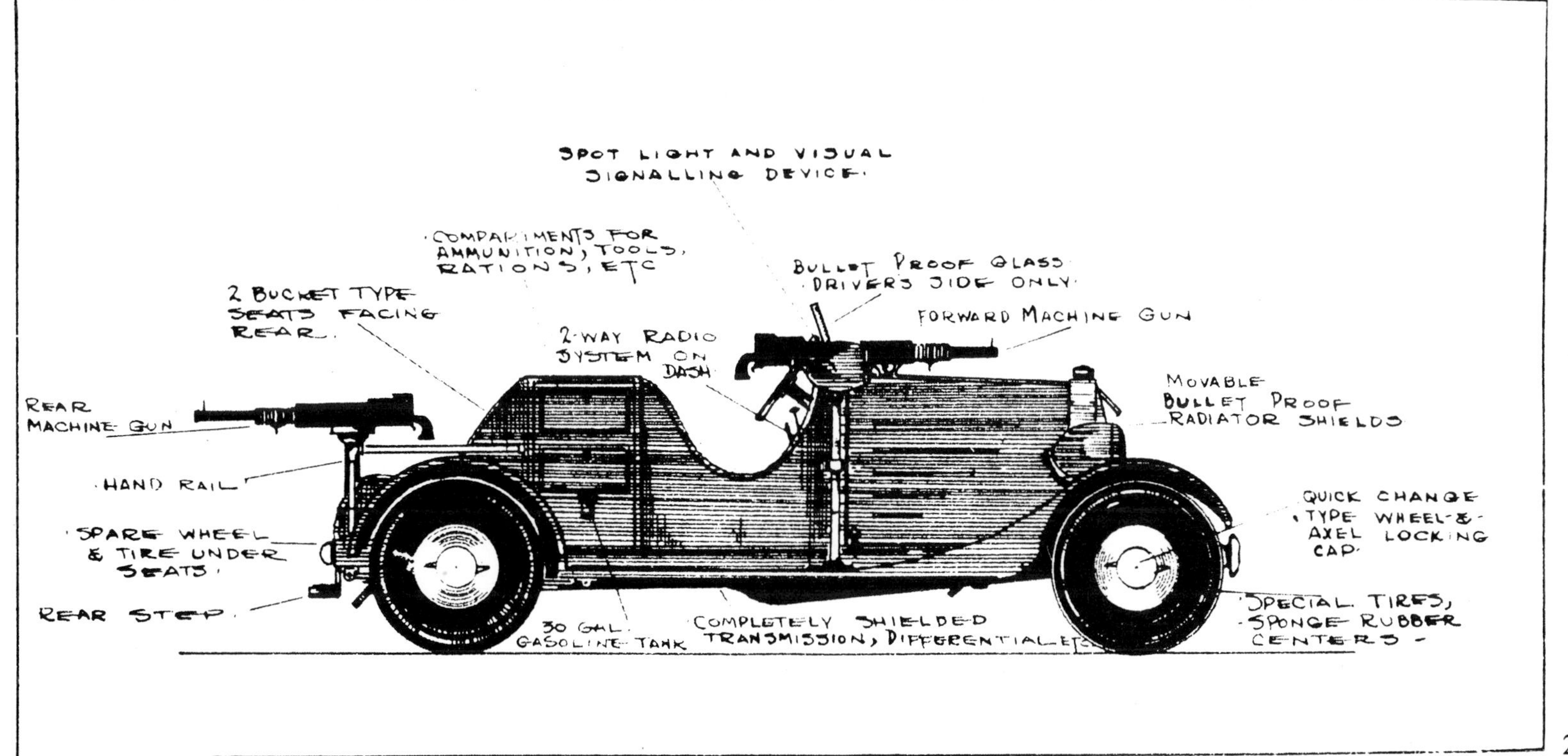

The technical services of the American Army were sufficiently interested to undertake a study of the idea and at the start of June 1940 they laid down the first requirements for the project. The general staff then handed the responsibility for the follow-through to the Ordnance Technical Committee. This inter-departmental unit appointed a sub committee to deal with the project. Its job was to formulate the requirements and draw up a complete specification for the vehicle.

The sub committee began work on June 19, 1940 by meeting at the Bantam factory at Butler to have a discussion with the management and also to form an opinion on the production capacity of the plant.

Major Howie, father of the *"Belly Flopper"* joined the delegation. Engineers from Spicer, the company specialising in transmission components also took part to help with their technical expertise in the field of 4 × 4, because it had been decided by now that any future vehicle must have two driven axles. Three military engineers, Bob Brown, Bill Burgan and W. Beasley decided on the main features of the vehicle the Army needed; four-wheel-drive, crew of three, arament one 0.30 (7.62 mm) machine gun on a monopod mounting, minimum speed 3 m.p.h. (5 km/h) at the most, ground clearance 6.5 in. (16 cm).

During a meeting in Washington on May 27, 1940, the Ordnance Technical Committee approved the final recommendations establishing the specification for the light 4 × 4 quarter-ton vehicle which the Army was hoping to order. It specified a maximum weight of 1,300 lb (590 kg) and a payload of 600 lb (272 kg). Maximum permitted wheelbase and track were 80 in. (2.032 m) and 47 in. (1.194 m) respectively.

Those submitting designs were required to deliver 70 vehicles, including eight with four-wheel steering, within 75 days. This was a strict and incredibly tight time limit but for good measure a first prototype had to be presented within 49 days, the balance of production to be delivered within the next 26 days. A budget of 175,000 dollars was allocated to finance the programme.

The Quarter Master Corps invited 135 manufacturers to submit proposals meeting these conditions. The procedure was remarkable not only for the importance of the manufacturers who were contacted, but for the number of replies received. In fact just two companies, the American Bantam Co. and Willys Overland Inc. replied to the invitation and sent in their proposals. It must be admitted that certain aspects of the project, especially the short time limit and the low maximum weight must have cast serious doubts on its feasibility.

As soon as the invitation to tender arrived at Bantam, Francis H. Fenn, the President of the company, sought the advice of Arthur Brandt who had been his predecessor at the head of the company in the days when it was still known as the American Austin Car Co. Brandt advised him to contact Karl K. Probst, a highly qualified engineer with whom he had worked in the past. At first Probst refused the proposition which Fenn made to him but on Wednesday July 17 1940 he finally agreed to manage the project on behalf of Bantam.

The tight time limit and the limited financial, technical and human resources available at Butler were doubtless factors accounting for Probst's hesitation. Nevertheless he got to work immediately. Every minute counted, because the written proposals had to be submitted to Holabird, the test centre for wheeled vehicles for the US Army, by 9 am. on the following Monday, July 22 at the latest. Probst rose to the challenge and in less than five days he drew up the plans for what was to become the Jeep. He got into touch with Spicer at Toledo to ensure supplies of axles and transfer gearboxes. The axle he chose was one made for the Studebaker Champion car, which weighed 2,090 lb (950 kg). With the help of Bob Lewis, a Spicer engineer, he quickly came to the conclusion that all the technical requirements laid down by the Quarter Master could be met with one exception, the weight. However, this was a problem that did not worry him too much because he was utterly convinced that no other manufacturer was in a position to meet the specification on this point.

For the engine, Probst decided on a Continental Y-4112 unit which was immediately available and had what seemed to be sufficient power (48 b.h.p. at 3,250 r.p.m.).

Harold Crist, Works Director of Bantam and Francis Fenn helped Probst to find other parts suitable for the new vehicle among those which the company was already making.

On Sunday September 21 the dossier was completed and Probst arrived in Baltimore to show it to Charles Payne who was responsible for Bantam's military sales. He expressed considerable doubts about the success of the enterprise when he learned that the estimated weight of the vehicle calculated by Probst was 1,850 lb (840 kg) which exceeded the specified figure by about 550 lb (250 kg).

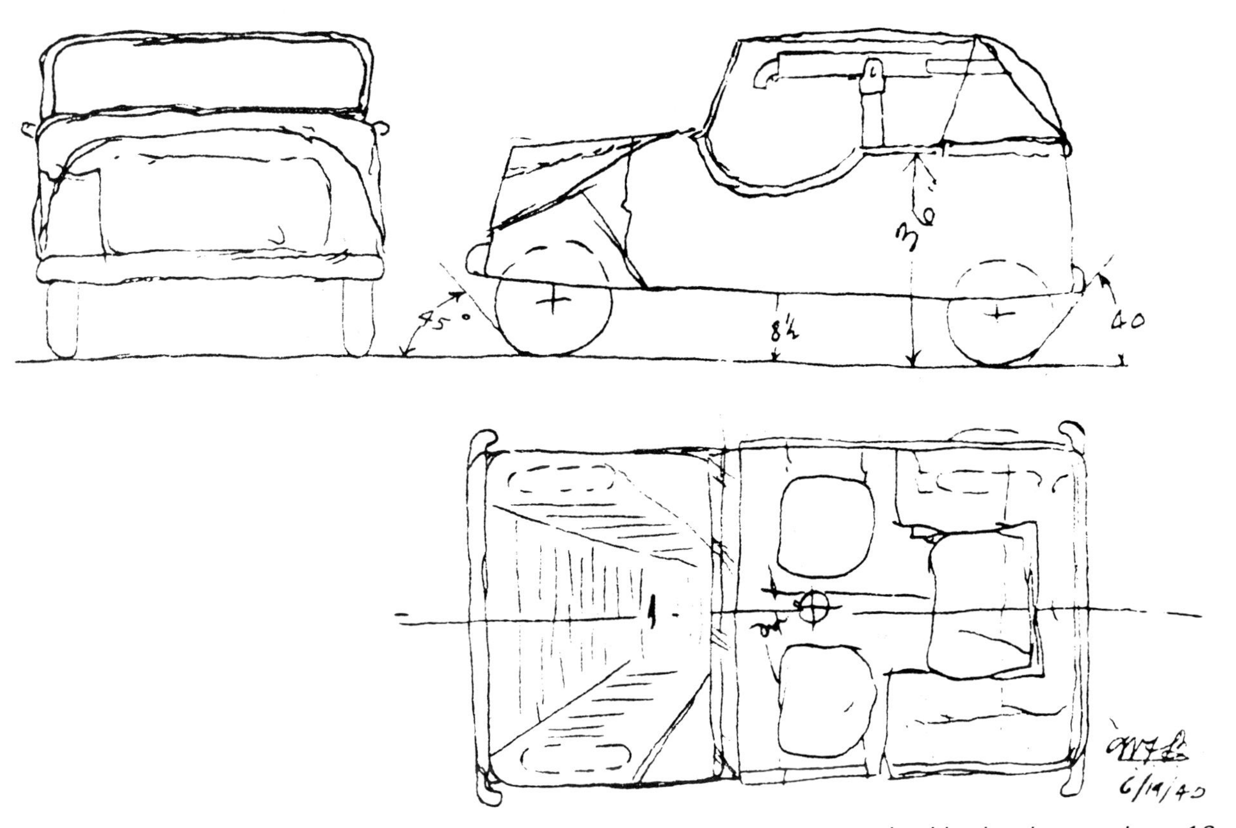

William F. Beasley, Chief Engineer of the US Ordnance Department, made this drawing on June 19 1940. It indicates all the basic characteristics of the future Jeep.

Payne also pointed out that the presentation was not compiled on the regulation forms, and so in the middle of the night, only a few hours from the time limit for the submission of proposals, a typist was brought in to re-type the entire document on the correct forms. They also took advantage of the re-typing to insert in the vehicle description the weight as specified by the Army.

On July 22 1940 the tenders were opened at Holabird Camp by Major Herbert J. Lawes. Whereas Bantam was prepared to gamble on completing the programme in the 75 days specified, Willys, who feared problems in obtaining axles from Spicer of Toledo, the only company which was then able to produce these components, requested an extension of the time limit to 120 days.

The price proposed by Willys was lower than that quoted by Bantam, but the penalty clause which raised the price by five dollars for every day's lateness above 75 days, brought the total cost of the Willys project out higher than that of the Bantam and so Bantam got the order with a quotation of 171,185 dollars. They were not officially notified of it until August 5, 1940 but the Bantam management, confident that they would win, had already begun recruiting personnel and had engaged four extra engineers to work on the project. Probst and his team lost no time in tackling the design of the first prototype, delivery of which was due on September 23 1940 at 5 pm. precisely.

In fact the prototype was ready on September 21 and Probst was able to use the two days thus gained in carrying out preliminary tests which extended over some 150 miles (240 km). Then on September 23

The seventh pre-production Bantam, now in the collection of the Smithsonian Institute. This was one of the seventy examples ordered from Bantam in winter 1940 in time to take part in the first mechanised maneuvers of the US Army in the following spring. Eight of them had four-wheel drive and steering.

Probst took the wheel and with Crist as passenger, left Butler for Holabird, about 170 miles (270 km) away, where he arrived at about 4.30 pm.

The tests at Camp Holabird went on from September 27 to October 16 covering nearly 3,500 miles (5,500 km) of which more than 3,000 (5,000 km) were over the difficult test courses of the camp either across country or over all types of road surface.

Willys and Ford sent observers down in October to examine the vehicle and they followed the progress of the tests with close attention.

When the Quarter Master was reproached with having permitted such behaviour he retorted that the prototype was government property and consequently he could allow anyone he wished to examine it, even other manufacturers who were competitors of the one who had designed the vehicle. In fact the reason why the QMC showed such insistence on extending these extraordinary facilities to Willys and Ford was that they believed Bantam would be absolutely incapable of undertaking a massive production order if the time came to issue one and therefore it was best to make sure in good time that they could rely on the collaboration of manufacturers of the size of Willys or still more of Ford.

At the conclusion of the trials differing opinions were expressed on the Bantam prototype.

Some, like Colonel R. Robins felt that the Bantam Jeep was too high and was short of power. The most frequent criticisms concerned its excessive weight and the necessity for frequent maintenance, but taking them as a whole the assessments were fairly favourable and Captain E. Mosely, one of those responsible for the trials, declared that the Bantam prototype was the best they had ever tested at Holabird.

The final report on the trials listed some twenty defects or mechanical breakdowns which was really very few for a machine put together so quickly; it concluded with a favourable assessment and declared that *"the vehicle had a good power output and met the requirements of the service."*

Bantam noted the points of criticism in order to put them right and start production of the other models as provided for in the invitation to submit proposals. Delivery of the whole batch of 70 vehicles including eight with four-wheel steering was completed on December 17, 1940 but even before all 70 of this pre-production batch had been delivered, the US Army decided to place a firm order for 1,500 on the

A historical photograph of the first Bantam prototype taken at the factory in late summer 1940 with all those responsible for its success including Harold Crist, Director of the company and Karl Probst himself.

The 62nd pre-production Bantam No. W 2015384 returning to Camp Holabird at the end of a day's testing during the winter of 1940-1941.

While crossing streams during its trials the Bantam was often stopped by ignition troubles. This photo was taken in spring 1941.

strength of the first test report. These were given the code name 40 BRC and received registration numbers from W 2018932 to W 2020431. However, the placing of this contract by the QMC raised new problems. Some senior members suggested that the order for 1,500 vehicles should be divided equally between Bantam, Ford and Willys. The motor transport subcommittee of the Quarter Master Corps agreed and on October 18 recommended the adoption of this procedure.

Charles Payne, the Bantam representative protested vigorously and appealed to Secretary of State for War, upholding his company's rights and making it clear that *"only Bantam had genuinely worked on the Jeep project and met all the requirements, and if at some time the future the company at Butler was not capable of fulfilling all the orders it was quite ready to provide drawings and collaborate with other companies to ensure production on a very large scale."*

In the end the general staff and Secretary of State Henry Stimson decided that this highly controversial first order should go entirely to Bantam. But the QMC did not accept defeat and continued to let it be known that it still had its doubts.

During this time Willys, who had taken their exclusion from the competition very badly, had gone ahead at their own expense, and incidentally on the advice of Major J. Lord, designing and building two prototypes of their own, one conventional and the other with four-wheel steering.

On November 14 the Quarter Master finally got his way. At the suggestion of Lt. Colonel Henry S. Aurand who was in charge of the quarter-ton 4 × 4 programme and with the approval of General Knudsen, it was decided that each of the three competitors should receive an order for 1,500 vehicles. One condition was imposed; the prototype must meet the Army specifications.

It is interesting to note that at this time Senator R. Reynolds was energetically supporting the idea of fitting all these vehicles with four-wheel steering which would make them more maneuverable but at the same time make driving more difficult and complicate the maintenance. He wanted at least a large proportion of the vehicles to be built this way and wanted Bantam to do it. The senior officers of the Cavalry agreed with him but the standardisation of the final model prevented Bantam from succeeding with the project in spite of its political support. On November 11, 1940 the model with two-wheel steering designed by the Willys, Vice-President of Engineering, Delmar J. Roos arrived at Holabird and trials began on the 13th. Ford had still not produced a prototype and that from Willys had only just begun its trials. At last, on November 23, 1940 Ford delivered its prototype, called the Pygmy to Holabird.

A Bantam 40 BRC during disembarkation tests.

The two new vehicles bore an astonishing resemblance to the Bantam Jeep. Both competitors had had the chance to study the Bantam prototype at leisure and they had very largely been influenced by it.

Comparative tests then revealed the qualities and defects inherent in each of the three models. It emerged that the Bantam Jeep offered a lower fuel consumption, while the Ford prototype had better steering and was more comfortable. The Ford was

The 40 BRC was built very quickly in order to take part in the motorised maneuvers of the US Army in Carolina in spring 1941. The 12,7 mm machine gun is on a ground tripod as no special mounting had yet been devised.

however handicapped by a lack of power from its engine which came from the Ferguson Dearborn tractor and a gearbox with only three forward speeds which came from the Ford Model A.

The Willys vehicle, which was called the Quad, proved more powerful and had a better performance over the test course. Colonel C.C. Duell, Army liaison officer at Holabird reported that *"The Willys engine out-classed all the others in power and produced a brilliant performance."* One problem remained however, because in spite of the decision of the QMC to raise the minimum weight from 1,300 lb (530 kg) to 2,160 lb (980 kg) the Willys Quad was still much too heavy at 2,423 lb (1,099 kg).

The US military authorities therefore found themselves on the horns of a dilemma, because the vehicle

The Willys Quad as it appeared when delivered to Holabird in November 1940.

Handicapped by an engine from an agricultural tractor, the Ford Pygmy, did not achieve a very impressive performance during the test programme at Holabird. Nevertheless it contributed its front end to the eventual design for the Jeep. Photograph taken in winter 1940-1941.

which in their view offered the most advantages did not comply with the official specification. It took a decision by the Under Secretary of State for War, Mr. Patterson, to untangle the situation by deciding that Willys should get a firm order for 1,500 Quads in spite of their weight.

Willys were entitled to feel very satisfied with the turn of events. Ward Canaday, one of their Vice Presidents was conscious of it and realised that he had won the first round but that the future of the vehicle and the possibility of more and larger orders depended on finding a solution to the weight problem.

Willys now had a choice; either they most design a new prototype with a lighter and less powerful engine or retain their Go Devil engine and try to save weight in other directions. Barney Roos opted for the second solution and every part of the car was re-examined.

Colonel Duell wrote *"In order to retain the Willys engine, Barney Roos saved weight by other changes, even in the paint. The result was so close to the limit that even a little dust or mud could have taken the vehicle over the regulation weight."* Roos was congratualted on his performance by General Anthony C. McAuliffe who was later to distinguish himself at Bastogne and by Major Lewis. This improved vehicle which had clearly *"borrowed"* its radiator grille from the Pygmy was given the name of Willys MA Command Reconnaissance. After the purchase of the 1,500 pre-production models from each of the three companies, (Ford GP, Bantam 40 BRC, Willys MA) the definitive specification for the Jeep was drawn up for the final trials organised by the QMC with the collaboration of the commands of the infantry, artillery, cavalry and armoured forces.

The majority of these vehicles never entered into service with the US Army but were sent to England and Russia in the end under the Lend Lease programme. Taking into account supplementary orders, the total numbers built were 2,675 Bantam 40 BRC, 3,650 Ford GP, and only 1,500 Willys of which 49 had four-wheel drive and four-wheel steering.

This Willys Quad still carries a civilian license plate in November 1940. One Quad had four-wheel drive and steering.

A Willys Quad in open country in the winter of 1940-1941. The excellence of its engine had already given it a performance superior to those of its competitors.

The design requirements were as follows :

- Maximum speed at least 55 m.p.h. (88.5 km/h).
- Minimum speed not over 3 m.p.h. (5 km/h).
- Fording capability 18 in. (45 cm).
- Possibility of fitting chains to the wheels.
- Maximum weight 2,100 lb (953 kg) for the conventional version and 2,175 lb (987 kg) for the version with four-wheel steering.
- Carrying capacity 800 lb (363 kg).
- Approach angle 45°.
- Departure angle 35°.

The final comparative tests between the three models competing for the main order then began. The first impressions formed during the previous tests were confirmed. Thanks to its more powerful engine, the Willys Jeep was definitely the best. In an article entitled *"The Jeep in Action"* which appeared in the Army Ordnance magazine for September 1944, Major E.P. Hogan wrote *"On the basis the first prototypes presented by Bantam, Ford and Willys which were tested and tested again at Camp Holabird in the course of long trials which were among the toughest ever carried out in this country, and finally on the basis of the trials carried out on the 1,500 examples of each series, the US Army made its choice in favour of Willys, whose Jeep came closest to meeting the specifications and the requirements of the Army. Moreover the price quoted by Willys was the lowest."*

Bantam had definitely lost the match but at the last moment Ford, backed up by the QMC in a manner which was astonishing, to say the least, very nearly landed the contract for 16,000 units.

However, the Office of Production Management (OPM) vigorously opposed this course of action, arguing that such a procedure would be totally contrary to the rules governing the award of contracts. Furthermore, Willys was the most competitive from

(continued p. 40)

Willys looked after their public relations right from the start. Here is Senator Mead driving a Quad up the steps of the Capitol at Washington in February 1941. Beside him is Irving "Red" Hausman, Chief test driver at Willys who had driven the vehicle from the prototype workshop in Toledo to Holabird.

The MA. With the exception of the front end, this is the Jeep in practically its definitive form.

During the last months before Pearl Harbor the MA rolls off the Willys assembly lines alongside saloons destined for the civilian market.

Four Bantams cross a ford during the maneuvers in the spring of 1941.

The US Army kept hardly any Bantams or Fords for its own use. The Russians and the British received the majority and the British had already made several modifications to this one.

A drawing made from a poor amateur photograph showing a 40 BRC modified by a military workshop in Alaska with better protection against mud and snow (1941).

The British used the 40 BRC as a reconnaissance vehicle with the Airborne Division and in the desert. The rather complicated mounting for the Bren machine gun enables it to be used as an anti-aircraft weapon.

A Ford GP flying high at Fort Sam Houston, Texas in summer, 1941. The typical US Army felt hat was abandoned a few months later. The number W 2017488 shows that this is the 66th GP built by Ford in 1941, part of the first batch ordered by the US Army (Contract W.398 GM 8887).

A 40 BRC of the British 6th Armoured Division identified by its white armoured fist on a black background.

The "owner" of this GP has devised his own rustic but effective protection against the cold at Camp Pershing, Iceland in 1942.

Edsel Ford driving General Borresteel in a GP which has just left the production line.

A GP of the 6th Armoured Division pulling a plough in England. See a cooling system expansion tank in front.

the financial standpoint with a price of $ 739 per vehicle and promised the quickest delivery. So in the end a contract was signed between Willys and the QMC on July 23, 1941.

The Under Secretary of State for War recalled *"It was General Knudsen who decided which company would get the order. And although the Quarter Master would have preferred that Ford got the order because of their production capacity, General Knudsen maintained that Willys ought to have it because they were the cheapest. And that is what happened."*

Immediately after the contract for 16,000 units was signed a meeting was held at Camp Holabird to consider what improvements could be made to the Willys Jeep Type MA.

Numerous, but relatively minor modifications were decided upon, many of them involving replacement of parts by standard units already fitted on other US

The first 25,808 Willys MB in production form were recognisable by the radiator grille, of iron bars.

Army vehicles. Thus a new air filter conforming with Bureau of Standards specifications was adopted and a 6 volt 40 A generator was installed because it was the standard QMC type already used on many trucks.

The 10 US gallon (37.9 L) fuel tank was replaced by one of 15 gallons (56.8 L).

A standard battery (Type 2H 6 volt) took the place of the civilian battery fitted on the first series of Jeeps.

The handbrake was moved from the left of the driver to the centre of the vehicle.

A double hoop was adopted for the folding top in preference to the single hoop fitted originally because it provided more space underneath.

The army asked for an axe and a spade to be provided on the left side of the vehicle.

Standard combat lighting with blackout lamps was installed.

Other modifications affected the steering column, the brake circuit and the dampers.

Some of these modifications were decided upon when series production had already begun. Thus the wheels and tyres had to be changed. A holder for a 5 gallon (18.9 L) Jerrican was fitted to the rear of the vehicle and a supplementary black-out lamp was mounted on the left front wing. A screened electrical system was also provided.

The model incorporating these improvements was standardised under the name of Willys MB. 361,349 of them were built, of which the first 25,808 examples differed slightly from those produced subsequently; iron radiator grille, shape of the fuel tank, instrument panel etc.

In October 1941, the need for Jeeps seemed so great that it was necessary to look for a second source of supply. An agreement was made under which Willys would supply their drawings to another manufacturer. On November 10, 1941, Ford was designated as the second supplier and a contract worth $14.6 million was placed with them for the supply of 15,000 Jeeps. In all Ford built 277,896 GPWs (General Purpose Willys).

Bantam could well have been brought into this expansion of production, especially as careful investigations made after the war confirmed that they had an industrial infrastructure which would have made them perfectly capable of producing Jeeps in large quantities.

However, the QMC continued to resist the idea and despite all the efforts of the company and its President, Bantam never obtained another order for motor vehicles. The company at Butler was confined

Although most of the MAs were sent to the USSR, a small number remained in the United States for use in training camps. The helmet and dress of driver and passengers indicate that this American photograph was taken in the second half of 1942.

to the production of trailers and aeronautical equipment for the whole duration of the war. Although it played such a large part in the conception of the Jeep, Bantam never got any benefit from it. Even worse, when peace was restored, Bantam went broke and in 1956 it was absorbed by American Rolling Mills.

On the other hand, Willys showed no hesitation in making unrestrained use of the popularity of the Jeep and its reputation for solidity in their own advertising campaigns. In fact the way they presented it was so shameless and improper that a law suit resulted in 1948. The Federal Trade Commission ordered Willys Overland Motors to cease the advertising in which they declared that the Jeep had purely and simply been invented by Willys in collaboration with the Quarter Master Corps.

The judgement recalled that Willys had neither invented nor created the Jeep and that Willys was not the only designer of it. The original idea for the Jeep must rather be credited to Bantam in collaboration

The first batch of Jeeps had no rear mounting for a Jerrican. The name Willys was pressed into the rear panel. The shape of the fuel tank was more angular.

Front of the same vehicle. Canvas side flaps give symbolic protection against mud. The number SM 2275 is the contract number (Supply Mechanical) for Lend-Lease under which the vehicles were supplied to the British.

with certain officers of the US Army and the vehicle offered by Willys on the civil market was not the same as those produced during the war.

In short, the Federal Trade Commission accused Willys of misleading publicity and unfair competition.

The difficult and controversial genesis of the 4 × 4 quarter-ton Jeep which had begun seven years earlier, thus ended in an inglorious fashion in a law court, but these squalid political and commercial considerations could in no way tarnish the intrinsic merits of the vehicle. The case made clear the rights of those concerned but it did not prevent Willys pulling their chestnuts out of the fire where the Jeep was concerned.

The fact is that the Toledo company was the only one to go on building Jeeps after the war and many people still believe that this was the company which invented the original concept of the vehicle. It is a very great injustice for Probst and his small team and for the technical services of the Quarter Master Corps without whom nothing would ever have been achieved.

Ford GPW used by the Marine Corps with their special radio equipment.

"The Russians soon appreciated the value of our Jeeps. They had asked for motorcycles and sidecars but as I said in a letter to Ambassador Litvinov at the end of January 1942, our Army now uses Jeeps almost exclusively instead of sidecars.

The Russians decided to try them and soon recognised that we were right. The Jeeps went so well in swamps and on the bad Russian roads that the Red Army soon demanded more. Since then we have sent 20,000 to Russia."

Edward Stettinius Jr. assigned by the President of the United States to administer the Lend-Lease programme reported this Soviet reaction in his book *"Lend-Lease : Arm of Victory"*. He went on to tell this story.

"Last year when an Associated Press correspondent went to see a Soviet artillery regiment on the central front he was taken in a Jeep through thick mud across churned up fields to the headquarters of the regiment. Between the bumps, he turned to the Red Army chauffeur and asked if he liked these rugged little cars. The chauffeur answered with a single word "Zamechatelno" which is the Russians equivalent of "Terrific!"

The Jeep quickly became famous among the public as well as the military. Its popular success was probably closely connected with the fact that war correspondents frequently mentioned the Jeep in their reports.

One of the most celebrated correspondents from across the Atlantic, Ernie Pyle who was later to die in a Jeep during the landing at Okinawa in April 1945 wrote of his Jeep *"I do not think we could continue the war without the Jeep. It does everything. It goes everywhere. It's as faithful as a dog, as strong as a mule and as agile as a goat. All the time it carries loads twice as heavy as those it was designed for and it keeps going just the same... The Jeep is a divine instrument of military locomotion."*

A Willys Jeep MB in its definitive form photographed at the factory on September 10 1943.

THE ORIGINS OF THE WORL

This is a subject which has been surrounded by controversy and the origin of the word Jeep is still very much an open question. Before the universally accepted name of Jeep was adopted with the success which has now surrounded it for forty years, other names were given to small quarter-ton 4 × 4 vehicles like *"Bug"*, *"Midget"*, *"Peep"*, *"Blitz Buggy"*, *"Quad"* etc.

The first prototype to be called a Jeep was the one built by Willys. Irving Hausmann the test driver of the Toledo company who drove the prototype to Holabird tells this story.

"Soon after I arrived, Ford also sent their prototype to be tested so we had to distinguish between our vehicle and the Ford because I was not keen on having any confusion between the two models. I therefore, launched the name Jeep among the soldiers who were there and at every opportunity I emphasised the name, which soon got around."

This version of the origin is confirmed by various military authorities at Holabird.

It therefore seems that although no single company can take credit for the paternity of the Jeep itself, Willys-Overland were the first company to use this name for their vehicle.

This name Jeep was taken up by the press and in February 1941 it was mentioned for the first time in the Daily News, a Washington newspaper.

The name was quickly adopted by the military and by the public and to this day although the name Jeep

JEEP

is a registered trade mark and the exclusive property of American Motors, in common language it continues to be used to describe any light vehicle with four-wheel drive and a general layout influenced more or less by the Willys of 1941.

It is worth noting that the name Jeep was given to other machines such as an agricultural tractor built to the Minneapolis Moline Power Implement Company which was later adapted for military use, also for an autogiro tested at Wright Field and for the prototype of the B 17 Flying Fortress tested at Langley Field in 1937.

The Dodge half-ton 4 × 4 was also called the Jeep in the beginning, a name which later gave place to *"Peep"*.

The etymological origin of the word Jeep has built around it diverse and controversial explanations. One thing is certain, a celebrated animal, a hero of the comic strips invented by an artist named Segar made its appearance in March 1936 in the adventures of Popeye. This animal, called Eugene the Jeep, was about as big as a dog, came from Africa, ate nothing but orchids and was able to make itself invisible. This wonderful little creature became very popular and by analogy anything astonishing soon came to be described as a Jeep.

For some people Jeep is an old army slang term going back to the first World War and applied without distinction to any new vehicle, or the name of a dance, or a name given by tramps to any odd-look-

"Eugene, the Jeep", an animal of very uncertain origin which became a friend of Popeye. Was this the origin of the word "Jeep"?

In the great tradition of the Western this cavalry sergeant prepares to give the coup de grace to his Jeep which has broken its "front leg". Bill Mauldin's sketch well expresses the feelings of drivers for their Jeeps.

ing individual. For others Jeep comes from the contraction of GP for *"General Purpose"*. (Webster's International Dictionary, Encyclopaedia Britannica) but although it is often quoted, this explanation seems doubtful.

An American writer, H.L. Mencken, author of a book on the *"American Language"* settles the question with humour. Against the word Jeep appears this question *"Jeep : can anybody give me the exact etymology and the history of the word?"*

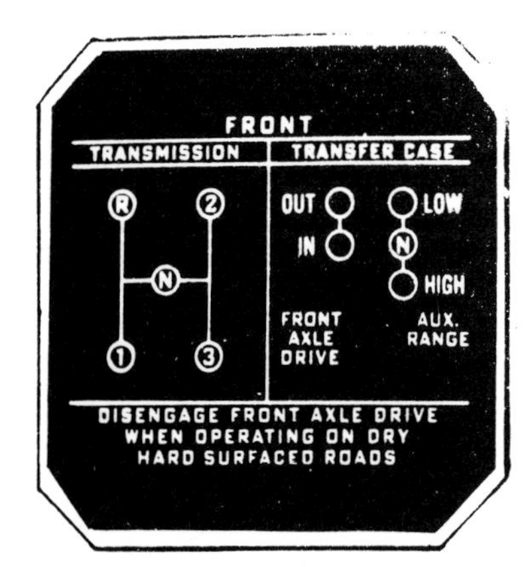

Gear shift diagrams and driving instructions for the Jeep were provided in four languages; English, Russian, Chinese and Spanish.

JEEP TECHNICAL FEATURES

One of the advantages of the Jeep is undoubtedly its simplicity and the standardisation of its components. Whether made by Ford or Willys, all the components apart from a few details are completely interchangeable. It is worth noting that Ford usually had the habit of stamping all the parts made by them with a capital F.

The Engine

Willys Type 441 or 442 generally known under the name of "Go Devil". It is a four-cylinder in-line water cooled with side valves, running on regular grade gasoline and developing 60 b.h.p. at 4,000 r.p.m. Swept volume is 2,199 c.c. (70.375 × 111.125 mm) No cylinder liners are fitted. The camshaft is chain driven except on the last series which had a gear drive. Compression ratio is 6.48 : 1. The plugs are either Champion QM 2 or Auto-Lite AN 7. The distributor is an Auto-Lite. There was a special distributor which was sealed against dust. The mechanical fuel pump is an AC model AF. The carburetor is a Carter Model WO 5395.

The differences between the Ford and Willys engine are minimal : they include the connecting rods, the crankcase breather, the inscriptions on the cylinder head and above all its method of attachment; entirely by studs on the Willys, whereas Ford use studs and bolts.

Transmission

Borg and Beck single dry plate clutch (model 11123); Warner gearbox Model T. 34J with three forward speeds and one reverse. Only second speed and the direct top are synchronised. The transfer box which is bolted directly to the gearbox can be a Brown-Lipe or a Spicer. It gives the choice of a high range in direct drive or a low range with a reduction of 1.97 : 1. A separate lever allows the front axle drive to be engaged when required. The low transfer gear range can only be used after engaging front-wheel drive. If desired a power take-off or various

Gearbox in	Transfer box in	
	Direct drive	Reduction gear
Top gear (2.665)	105 km/h	53 km/h
Second (1.564)	66 km/h	34 km/h
First (1.000)	38 km/h	19 km/h
Reverse (3.554)	29 km/h	14 km/h

Year of manufacture
Chassis numbers and years for the Willys MB

Year	First number of the year	Last number of the year
1941	MB 100 001	MB 108 5989
1942	MB 108 599	MB 200 022
1943	MB 200 023	MB 293 232
1944	MB 293 233	MB 402 334
1945	MB 402 335	MB 459 851

other systems can be fitted like for example the pulley which drives the bilge pump on the GPA.

Front and rear differentials are by Spicer as are also the axles. For the front axle, the constant speed universal joints may be by Bendix, Weiss, Rzeppa, Spicer or Tracta.

Chassis

This is a conventional ladder type with longerons and cross members made by Midland Steel. Although identical in layout, the Ford and Willys frames differ in detail. The cross member supporting the radiator is round on the Willys and an inverted U on the Ford.

Technical Characteristics according to TM — 9 — 2800 of September 25 1943

Willys MB or Ford GPW

Empty weight	1054 kg
Carrying capacity	363 kg
Gross weight	1417 kg
Crew	2 men
Number of seats	4
Tyres	4 single 600 × 16 6-ply plus a spare wheel.
Fuel tank	57 litres 70 octane gasoline
Brakes	Hydraulic
Lenght	3.360 m
Width	1.580 m
Height (to steering wheel, windshield folded)	1.320 m
Height, total	1.770 m
Wheelbase	2.04 m
Engine	Willys MB 4 cylinders in line or Ford GPW with governor (3,820 r.p.m.)
Maximum gradient	60 %
Approach angle	45°
Departure angle	35°
Turning circle	10.640 m
Fuel consumption on the road	11.7 lit/100 km without trailer
Maximum speed	105 km/h at 3,820 r.p.m. with governor

The inner face of the longeron is pierced by circular holes on the Ford and totally open on the Willys. The central plate supporting the body which forms the anchorage for the machine gun mounting is rounded on the Willys and rectangular on the Ford. The Willys chassis has an additional cross member at the rear. Finally, the shock absorber mountings are bell shaped on the Ford and L shaped on the Willys, but in spite of these small differences, all the attachment points are strictly in the same places.

The leaf springs, identical on right and left at the rear, have nine leaves. The right front spring has eight leaves and the left one is identical but has a supplementary half leaf to compensate for the extra weight due to the fact that the engine is mounted off centre and because vehicle is often driven with only the driver on board. The telescopic dampers are hydraulic, those at the rear being reinforced.

Braking is hydraulic, with drums on all four wheels, all the components coming from Bendix. One of the faults on the Jeep was the inefficiency of the handbrake. Only its simplicity had justified its adoption in production. Therefore on the later models of transfer gearbox built during the war the handbrake was improved by replacing the brake band with a drum brake. This was also carried out as a retrospective modification in the workshops of the American Army after the war. Although the responsibility for this modification is generally credited to Ford, it was fitted as standard by both Willys and Ford. Hotchkiss adopted this improvement for their handbrake.

The first models had wheels with rims in one piece but in mass production they were generally fitted with detachable combat rims consisting of two flanges. The tires were 600 × 16 6-ply cross country with a military tread pattern.

Electrical Equipment

Six-volt electrical equipment with negative earth. Headlamps are sealed beam white light units. The early headlamp switches were push-pull; later they were replaced by a rotary switch.

Dynamo, starter and voltage regulator are by Auto-Lite. After the war the US Army converted its Jeeps to 12 or 24 volts. Incidentally, Ford introduced the idea of a swinging mounting for the sealed beam headlamps which allowed them to illuminate the engine compartment. This feature was then unique on military vehicules and facilitated repairs and adjustments at night.

A Bantam trailer being put to good use.

Quarter-Ton Two-wheel Cargo Trailer

As an auxiliary for the Jeep a trailer was designed to double its carrying capacity. With a cross country capacity of 500 lb or 226 kg and a capacity of 453 kg on normal roads it was classed as a quarter-ton trailer. It could float with a full load, its ribbed steel panels forming a bouyancy compartment. This theoretical advantage was rarely utilised in practice and made loading difficult because none of the faces of the trailer could be dropped down.

During the war many Jeep trailers had their rear panels cut out and fitted with hinges by the people using them. The Model M 100 which took its place after the war looked identical from the outside but was soon fitted with a drop-down tailgate.

A tarpaulin was provided to protect the load against the weather. Manufacture of the trailers was the consolation prize awarded to Bantam who built more than 100,000 of them during the war under the name American Bantam T3 and went on building them until 1952.

Willys also built two-wheel trailers for the Jeep under the name of MBT (MB Trailer). Many other manufacturers also took part in the programme including Fruehauf, Gemco, Springfield, Pacific, Black, Checker and Strick.

	Trailer 1/4 ton Bantam Willys	Trailer 1/4 ton M 100
Floatable	Yes	No
Length	2,794 m	2,786 m
Width	1,422 m	1,422 m
Height	1,016 m	1,066 m
Brakes	Mechanical	Mechanical
Weight, empty	249 kg	261 kg

Two Scottish soldiers waterproofing the engine compartment of their Jeep. While one instals the flexible air pipes the other protects the plugs with Abestos from the pot on the wing (April 1944).

(Right) Drawing showing the special rigid exhaust extension for the Jeep. The waterproofing procedure described in the text was most widely used for all US Army wheeled vehicles. A; the curved upper exhaust pipe. D; attachment to the standard exhaust pipe.

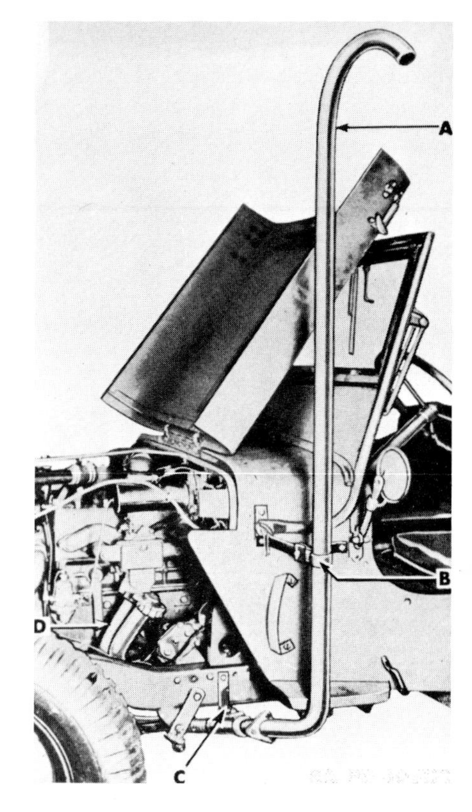

PREPARATIONS FOR USE IN DEEP WATER

Although the electrical system is not waterproof or even shielded, the Jeep can travel without difficulty in nearly 50 cm of water. Above that however, one is almost certain to be immobilised by a breakdown.

A waterproofing kit was therefore developed for use in deep water so that Jeeps could safely travel the last few yards separating the landing craft from the shore and move quickly to a position better sheltered from enemy fire. To waterproof a Jeep or any other similar vehicle (Dodge, GMC, Chevrolet) the following items must be protected against water.

- The whole electrical system, to protect the ignition and eliminate the risk of a short circuit.
- The supply system for air and fuel to ensure that the air/fuel mixture remains combustible.
- The exhaust system in order to prevent water rushing in if the engine stops.
- All lubricated components which by definition are vulnerable to water (engine, gearbox, axles).

After having cleaned, dried and checked all connections, the electrical circuit was protected with a red synthetic paint called Glyptal which had high insulating properties and it was then covered with a generous coating of an asbestos grease called Abestos.

The exhaust system was replaced by a flexible metallic pipe connected to the manifold and attached to the left upright of the windscreen. A similar pipe in reinforced rubber was attached to the carburetter in place of the air filter, passing under the bonnet and clipped to the right upright of the windscreen.

with a circular dome at the top to keep out drops of water.

Similar measures were taken to ventilate the distributor and the fuel tank with slim flexible rubber pipes.

After installing these various pipes, all the components vulnerable to water, that is to say the whole electrical system, battery, coil, voltage regulator, plugs, dynamo, starter, headlamps, horn, instrument panel, the whole oil and fuel circuits, the fillers and drain plugs for the gearboxes and axles as well as the axle shaft had to be covered with a thick coating of Abestos.

To complete the preparations the flexible exhaust pipe had to be covered with a fireproof paste and an insulating compound had to be sprayed over the whole engine compartment. Althought it was a simple programme it took up a great deal of time and attention an only protected the Jeep from the water for a maximum of eight minutes in a depth not exceeding three feet (1.4 m).

It is easy to understand why the tendency later was to equip military vehicles with watertight components and protected electrical circuits. This was the case with most of the Hotchkiss M 201 vehicles.

From the time it went into service the Jeep was constantly being tried out with various modifications either by the technical services of the US Army or the field workshops in order to fit it better for the many different tasks it was required to perform.

In preparation for the landings in Europe the Allies ran driving courses which included fording through deep water. This forced the crews to fit the Deep Fording Kit very carefully, something which they could not always be bothered to do. (February 1944).

The first Jeeps to disembark in Normandy photographed in the British sector on June 7 1944 have the Deep Fording Kit, but only the air inlet pipes are in position.

This Jeep of the 25th Marine Division which has had a hard time in jungle fighting still carries part of its Deep Fording Kit. The air inlet pipe has been attached directly to the carburetter through a hole in the bonnet.

ARMAMENT FOR THE JEEP

First attempts to arm the Jeep were carried out with Bantams at the maneuvers in Carolina during the Spring of 1941, the 12.7 mm machine gun being mounted on its normal tripod fixed to the standard floor of the vehicle. A 7.5 mm gun was also mounted on a fairly low column while retaining its ground tripod folded up underneath. These arrangements were not particularly convenient and new mountings for the Jeep were investigated.

In the end a monopod mounting for a 12.7 mm. (Type M 31) machine gun behind the front seats and a support for 7.62 mm machine gun (Type M 48) fixed to the instrument panel in front of the passenger became standard equipment for the Jeep. Many experiments were made, as listed below, usually aimed at increasing the fire power of the Jeep, but came to nothing.

1941 : Fitting of two 12.7 mm machine guns on monopod mounting for use against aircraft. These experiments were conducted by the Cavalry Board.

1942 : Navy Mark 21. 30-calibre machine gun on a

A Jeep used by the Spahis in the liberation of La Rochelle with the pedestal mounting for the M 31 12.7 machine gun developed specially for the Jeep.

A Jeep of the American Army in the South of France in August 1944 with a 12.7 machine gun on an M31 mounting. Clipped to the windscreen is an air inlet pipe for use in deep fording operations.

The second gun mounting for the Jeep was the M48 for 30-calibre machine guns. The photo shows how difficult it was to use. Most post-War models had an offset pivot for the gun which gave the gunner more elbow room. (Alsace 1944).

raised circular mounting above the front passenger seat for use as an anti-aircraft weapon.

1942 : Navy Mk 27. Mounting for a 30-calibre machine gun on the instrument panel on the right.

- Gun mounting with pivot turning the 30 calibre machine gun to the right. This mounting received no order number and is identified by the drawing number, D.76272.

1944 : T 45. Rack for 14 rockets fixed on the sides of the Jeep. This mounting used by Marines in the Pacific was more usually carried on an International.

- Conversion of the Jeep in Europe with rocket launchers. The front compartment and windscreen were protected by plating. Two rows of six tubes. Used mainly during the campaign in Alsace.
- 4.2 in. mortar with bipod mounted in the centre of the Jeep and the rear body panel removed to allow room for the base plate which rested on the ground when the weapon was fired.

1945 : T 21 recoilless 75 mm gun at the rear on the right. A tubular structure in front controlled the firing angles of the gun. Some tests were made with twin 75 SRs.

- T 19 recoilless 105 mm gun mounted in the middle of the vehicle.
- The Jeep was also the normal vehicle for towing the 37 mm anti-tank gun.

Obviously during military operations other mountings were devised by the troops on the spot and some ideas of them can be gained from the illustrations because there was such a variety, bearing in mind the tasks to be accomplished, the local resources available and the armement, which was either regulation or "borrowed" from other armies or even salvaged from damaged aircraft.

During the campaigns in Indo-China and Algeria, the French Army devised various adaptations for light automatic weapons (Browning BAR and 24/29 light machine gun, Frenc MAC machine guns and Americans 30s) besides a mounting on the right of the Jeep which allowed the mounting point of the gun to be folded down towards the front. This mounting which was used on US Jeeps as well as on the M 201 is much easier to use than the American model M 48 as the gunner has more space and freedom of movement with his gun moved forward.

A Jeep from a reconnaissance unit of the 2nd French Armoured Division in Alsace with a 7.5 machine gun on M48 mounting.

Various mountings for the 30-calibre machine gun were tried before the M 48 was standardised, including this Mark 27 devised for the US Navy.

The Mark 21 mounting developed by the US Navy was intended to provide a 360-degree field of fire against ground and air targets but was not adopted.

HOW TO DESTROY YOUR

In the American Army instruction book a complete chapter is devoted to the various methods of destroying the vehicle so as to prevent the enemy using it or obtaining spare parts from it. These lines of action are strictly defined and constitute a veritable bible of destruction.

"All parts essential for the use of the vehicle and all essential spare parts must be destroyed or damaged to the point where they cannot be repaired.

For the destruction, choose a place which will provide the maximum obstruction to enemy movements while creating the least risk for friendly troops, because during the destruction ammunition or debris may find targets in the neighbourhood. Always take the necessary safety precautions."

Methods of Destruction

a) Breaking up by mechanical means : use sledge-hammers, axes, hachets, picks, hammers, levers or any other heavy tool.

b) Gashing : use axes, hachets or machetes.

c) Fire : use gasoline, kerosene, flame throwers and incendiary grenades.

d) Explosives : use firearms, grenades or TNT.

e) Other methods : use anything immediately available to destroy the equipment.

f) Put it out of reach : bury in trenches or other holes, throw it into a river or canal, hide it or scatter it about.

Destruction of the Vehicle

Three methods of vehicle destruction are described below in their order of efficiency. Whichever method is used, it is essential to follow the order indicated to ensure uniformity of destruction in a group of similar vehicles.

Method 1

1) Unload and empty portable extinguisher, puncture fuel tanks.

2) Using an axe, a pick, a sledge-hammer or any other heavy object, smash all important components such as distributor, carburetor, air filter, generator, ignition coil, fuel pump, spark plugs, ignition equipment, instruments and controls. If you have time and something heavy enough, smash the cylinder block, the cylinder head, the crankcase, the gearbox and the axles. Slash and destroy the tires.

3) Splash gasoline and oil over the whole vehicle and set it on fire.

Method 2

1) Unload and empty the portable extinguisher; puncture the fuel tanks.
2) Fire on the vehicle with a mobile gun, a tank or artillery, rockets or anti-tank grenades. Concentrate on the engine compartment, the axles and the wheels. If a good fire breaks out the vehicle can be considered as destroyed.

Method 3

1) Unload and empty the portable extinguisher; puncture the fuel tanks.
2) Prepare two charges of four half-pound sticks of TNT or equivalent charges with a non-electric detonator and about 6 ft. of slow fuse for each two-pound charge. Place one charge on top of the clutch housing and the other as low as possible on the left side of the engine. Caution : if the charges are prepared in advance and carried in the vehicle, always separate the TNT charges, the detonators and the slow fuse until the moment when they are to be used.
3) Ignite the fuses for the TNT and take cover. The danger zone extends to about 200 yards. The slow fuse will burn for about 3 1/2 minutes.

Destroying the Tires

a) Attempts must always be made to destroy the tires including that on the spare wheel even if time does not permit the destruction of other parts of the vehicle.
b) Tires can be destroyed by incendiary grenades. Set off a grenade under each tire. If this method is combined with the destruction of the vehicle by TNT, the fire started by the incendiary grenades must be well established before the TNT charges are exploded.
c) Tires can be destroyed with an axe, a pick or by fire from a heavy machine gun. Where possible deflate the tires first. You can also pour gasoline over the tires and set fire to them.

It goes without saying that we have not found any illustrations of these methods of destruction at the front as they generally took too long to carry out, either because the troops kept their Jeeps in order to retire more quickly or because they solved the problem by setting fire to the vehicle or throwing a grenade into it. It is interesting moreover that the Germans were very keen on Jeeps and used them happily whenever they could capture one in good condition.
In fact intentional destruction was most often carried out to open up a road or clear an area littered with broken down vehicles and it was done in the quickest way by using a tank-dozer.

The T 28 was fitted with a Bombardier flexible half track. The front tyres were 7.50 × 16. Registration numbers beginning with 40 were allocated to US Army half-tracks.

JEEP SPECIALS

THE HALF — TRACK JEEPS

With the extension of the war and the Japanese attack on the Aleutians, Alaska became a military zone of primary importance. Intense aviation activity created enormous difficulties for personnel confronted with a cold climate and bad weather for the greater part of the year. This, quite apart from war operations resulted in many aeroplane crashes which were made particularly critical by natural forces, fog, cold, snow and ice.

The US Army Air Corps tried to provide its rescue services with vehicles that could cope with the terrain. As the main problem was the operation of rescue vehicles over vast areas covered by deep, soft snow, it seemed logical to turn to a company which specialised in building vehicles to travel in these regions. They got into touch with the Canadian Bombardier Company which had developed a series of tracked vehicles using rubber wheels of large diameter driving a wide, flexible track with metal cross bars. They developed a steering track system for the Jeep which again looked like the ideal vehicle for conversion. The first model, called the Half-Track Snow-Tractor T 28, took the form of a lengthened Jeep with an overall length of 4.06 m and a rear compartment completely redesigned to permit the fitting of flexible track by Bombardier or Chase.

The pneumatic tyres at the rear with cleats forming a driving wheel were 4.75 × 19; the other two tyres were 4.00 × 12. Oversize front wheels were fitted (7.50 × 16) and the output of the standard Go-Devil engine had been raised to 63 b.h.p. at 3,900 r.p.m. The two front tyres could be replaced by big skis when the snow was particularly soft and deep. Thus equipped, the machine had a ground loading of less than 1 kg/cm^2. As a result of these first tests, a new version, a little lighter, with a smaller rear compartment was tried out under the name T 29. Further modifications to the body led to the final version, the T 29 E 1. In spite of lengthy tests, the Half-Track Jeep never won approval and was never put into production. On the other hand the vehicle which was

The T 29 snow Tractor differed from the T 28 mainly in its rear bodywork. The leaf springs were fitted upside down.

The T 29 E 1. The front wheels could be replaced by skis for use in deep snow.

The T 29 E 1 on test in Alaska, a view which shows the different bodywork compared with the T 29. Note the angle of the steerable rear track in entering a corner.

For patrol work along the American coast, the US Coast Guard built this Jeep with a lengthened chassis called the Invader, which could carry 10 men. It had tyres of large section. The US Army carried out similar conversions.

The Willys ML W 4 was a prototype built in 1944 to carry a load of 500 kg. Wheelbase was lengthened by 30.5 cm. There were two versions differing in the types of axle and the tyres, either 7.50 × 20 or 7.50 × 16.

Throughout its career the question of giving the Jeep four-wheel steering cropped up. This is one of the last experiments of its kind. Although improving maneuverability it involved too much mechanical complication and made driving more difficult, something the Jeep did not need.

chosen instead of it was built by Allis Chalmers for the same job, and used the engine, clutch, gearbox, differential and steering of the Jeep.

Jeeps with Six-Wheel Drive

The Tank Destroyer Command which set up units of tank destroyers and also ran the procurement to provide equipment for them was the most recent branch of the American land army. It was independent of and was sometimes said to be antagonistic to the Armoured Divisions and the Cavalry Command.

As no equipment had been developed for its use before the war, it was largely reduced to improvising with the only anti-tank gun then available in the USA, the 37 mm. The performance of this weapon was rather limited, so the Tank Destroyer Command sought ways to improve its efficiency and mobility.

The 37 mm Gun on this six-wheeled T 2 E 1 could only fire to the rear, a feature of the first generation of tank destroyers tried by the US Army.

A Willys MT Tug as a tank destroyer armed with a 37 mm gun (Carriage, Motor 37 mm Gun, T 14). Six examples were built in 1942. The T 13 was identical but built on a Ford chassis.

With this object they tested the gun on a variety of wheeled vehicles.

They decided to concentrate on a six-wheeled Jeep but series production was not authorised because only Tank Destroyer Command was interested in this special chassis. Moreover its manufacture would have created problems for Willys who were having difficulty in producing their quota of standard Jeeps.

In the end only one model was adopted, standardised and produced in quantity; the Carriage Motor, 37 mm Gun M 6 which used the standard 3/4 ton Dodge chassis (see The Dodge in the same series).

The story of the six-wheeled Jeep is therefore tied up with that of the Tank Destroyer Command which was trying to provide itself with gun carriers sufficiently fast and maneuverable to compensate for the poor performance of its anti-tank gun. Being quick and maneuverable the machine could take advantage of the lie of the land or take cover to set up an ambush and hope to catch its prey at short range, thus compensating for the lack of punch in its gun.

This explains the position of the gun, which pointed rearwards, because as soon as it had been fired the vehicle had to get away quickly without further ado. The Tank Destroyer Command was therefore looking for a light, fast vehicle because the 37 mm gun and its ammunition were not very heavy and if possible something already available. The first tests were carried out with two Bantam 40 BRCs at the end of 1941. The first had been given the designation Gun Motor Carriage 37 mm T2 and was fairly extensively modified; the whole body at the rear had been removed and the steering column repositioned to allow the barrel of the 37 to revolve on its mounting without meeting any obstruction.

To achieve a centre of gravity as low as possible and an inconspicious profile, the gun was mounted very low on a pivot which obliged two men of the crew to get out to operate it. As explained later, this inevitably condemned the T2.

The T2 E 1 was at once less ambitiou and more practical because it consisted simply in mounting a 37 mm gun at the rear of a 40 BRC without altering the body. The gun had a limited lateral traverse and could only fire rearwards.

These first two attempts showed the interesting possibilities in using the Jeep as a tank destroyer but also its limitation from the point of view of carrying capacity as it was only able to carry a limited amount of ammunition and a crew of only two men.

37 mm anti-tank gun on a Bantam 40 BRC (Carrier, Motor 37 mm Gun, T 2). The steering column has been repositioned on this model photographed in 1941.

This last point alone comdemned the whole idea. Two men were needed to operate the 37 and reducing the crew to this number made it impossible to have a driver sitting ready to move off the moment the shot had been fired, an imperative requirement if the vehicle and its crew were to have any chance of survival. It must not be forgotten that the only armour provided was that of the shield on the gun itself which was not very thick and to have any effect the 37 must be used at short range.

To carry more ammunition and three or four men, something more ambitious was needed. To do this they thought of simply lengthening the Jeep chassis, moving back the rear axle and adding a third driven

T 24 Scout Car built by Willys with armour plating by Smart.

axle which was naturally accompanied by modifications to the suspension and the transfer box. This chassis, called the MT-TUG was at first given a body with a steel floor and side panels in wood.

The first model, called the Gun Motor Carriage 37 mm T 13 was based on a Ford GP chassis. Its gun pointed towards the front and it had a three man crew. The next type, the T 14 built in the early months of 1942, was constructed on a Willys MT-TUG chassis at the Aberdeen proving ground. This chassis had a wheelbase 51 cm longer (2.54 M) but retained the same engine and gearbox. The gun, pointing forwards, was protected by a rather rudimentary open topped box of armour plating. Once again there was a three man crew. The vehicle was 5.76 m long and weighed 1.290 kg.

Six prototypes of the T 14 were built and proved so satisfactory that Tank Destroyer Command asked for the model to be standardised for series production. Faced with objections raised by the ordnance services and the war production authorities who were not particularly anxious to impose a type of vehicle which was needed by only a single user on production lines which were already overloaded, various expedients were adopted to prove that the vehicle could fulfil other roles and so interest a greater number of units. To this end, a reconnaissance vehicle called the Scout Car T 24 was built, based on one of the Willys MT-TUG chassis with help from Smart on the armour plating. It was tested during the early months of 1942. Incidentally, the T 24 had a certain number of components in common with the T 25 which was to begin its tests in June of the same year.

The idea of the Tank Destroyer Command was a good and even audacious one. As the Armoured Board was interested in an armoured Jeep and as the first tests had shown that the normal Jeep could not handle the weight of the armour plating, why not take advantage of the 6 × 6 chassis to create a light reconnaissance vehicle which would then be a practical proposition?

The T 24 still had a three man crew with armament consisting of 12.7 mm machine gun on a monopod mounting. The machine weighed 2.472 kg. A project for an armoured troop carrier was also examined using the MT-TUG chassis and certain body parts from the T 24.

In spite of this, the chief of the General Staff of the US Army refused to authorise the adoption of the machine in any form whatsoever, preferring to rely for the mobility of the 37 mm anti-tank gun on a roomier vehicle on which the gun could be fixed with practically no modifications and moreover a vehicle which was already in production in large numbers for the US Army, the 3/4 ton Dodge.

The Armoured Jeeps

One of the jobs for which the Jeep had been designed was reconnaissance. It was therefore natural to think about providing it with light armour to give its occupants better protection on this kind of mission.

The Smart company of Detroit was the first to think of it. The company specialised in the construction of armoured vans for the transport of bank money and in building executive limousines for personages throughout the world who wanted to travel in luxurious automobiles without, at the same time neglecting security. First tests of the Smart armoured

T 25 Willys — Smart Scout Car on an MA chassis.

Jeep took place in November 1941 at Aberdeen, which had become the test centre for all vehicles proposed or manufacturered for the US Army.

The tests were conducted by the Armoured Board. The study services of the U.S. Army were at that time divided into numerous technical committees, each on responsible for the projects for the arm with which it was concerned. This quickly led to a fine mix up, the more so because cach committee was a law unto itself and could propose the adoption of models which suited it without bothering about what its neighbours were doing. The Armoured Board was the authority for all armoured conbat vehicles with the exception of tank destroyers which were the concern of another committee.

The armoured Jeep based on an MA proved disappointing because the suspension could not handle the additional weight of the armour plating, although the armour was very modest, consisting merely of steel plates 6 mm thick replacing the bonnet, the radiator grille and the windscreen, which could be folded down as usual. Besides driving vision slots which could be closed by sliding plates, two firing apertures were provided in the windscreen. They were circular and allowed the crew to use their firearms while protected by the armour plating, which had the same area as the original windscreen. The firing ports were a design of the Smart company which used them on all its vehicles. The prototype was taken back by the manufacturer who lightened it and added two half doors in armour plate.

The vehicle was ready to resume testing in June 1942. On this occasion it was given an official identification in the US Army series for reconnaissance vehicle prototypes : T 25 Scout Car.

The vehicle, still based on an MA chassis, was later completely redesigned by Smart, again with the object of getting the weight down to the minimum. The new version was called the T 25 El- the letter E indicating that a vehicle had been modified and the figure showing the number of times it had been modified. It now had protection for the front of the vehicle, and a windscreen and two doors all of the same height. Compared with the T 25 the windscreen was now fixed and smaller, as it only came level with the eyes of the front passengers and the firing ports had been eliminated. The protection was concentrated entirely on the front of the vehicle and there was no protection at the rear.

Smart would not give up and, still in 1942, began experimenting with an MB chassis leading to the

In 1941 the Smart Company, specialists in armour plating for official cars and bank security vans, suggested an armoured version of the MA. The first version, the T 25, had protection for the engine and a windscreen of armour plate.
The third, seen here, was the much more elaborate T 25 E 1 but the weight ruled it out.
However, the idea of fitting armour to the Jeep did not stop there and there were many improvisations in this direction, the first apparently on the Tunisian front in 1942.

T 25 E2 which had armour plating surrounding the whole vehicle thus protecting the rear passengers as well, the theoretical crew of the vehicle being three persons. The extra weight was then 447 kg.

On the T 25 E3 which followed it in the same year, the general shape of the armoured body was improved by replacing the vertical panels of the T 25 E2 with panels slightly sloped in V form which increased the ballistic resistance of the armour plating and the front surface was provided with three openable vision slits which at last allowed the crew to see something, because the two slits in the previous prototype limited visibility far too severely.

The T 25 E3 therefore turned out satisfactorily in many respects but the weight problem had still not been solved; for from it, because the vehicle was now carrying 465 kg of armour plating which was more than any of its predecessors.

This led to the tests being abandoned in 1943. Although the learned specialists of the Armoured Board never took the Smart experiments too seriously because the Jeep *"simply could not support such a weight"* their opinion was doubtless never made known to users of the vehicle who did not hesitate to fit their Jeeps with armour plating.

Smart had the right idea in spotting the need and the many improvisations made in all theatres of operation by nearly all the belligerents bear witness to the fact.

Lightweight Jeeps

As we have seen, the Jeep kept putting on weight in spite of all the efforts to control it. But some uses, mountain and jungle campaigns for example involving difficult terrain, required light, compact vehicles which could still carry a useful load of 250 kg with two men on board.

Finally airborne troops were widely used in the course of the conflict and they demanded light, compact equipment which could easily be transported in aeroplanes or gliders.

Crosley, Chevrolet, Kaiser and Willys built a number of prototypes in the United States but nothing came of them, either because the weight reduction they had undergone made them fragile and limited their performance or because the carrying capacity and the interior dimensions of aircraft had progressed so well that they had become unnecessary.

Crosley showed the first model in February 1943. After tests at Ford Benning in Georgia, 36 examples

The Willys MB-L special truck quarter ton-light.

Kaiser Midjet Jeep. It weighed 726 kg fully laden. The engine was a four-cylinder 30 b.h.p. air-cooled Continental.

of an improved model, the CT 13 PUP were ordered. Weighing 738 kg ready to roll, it was driven by a 13 b.h.p. 2-cylinder air-cooled Waukesha engine.

After the CT 13, which was not successful although six examples were sent to various theatres of operation, Chevrolet entered the fray with a vehicle weighing 708 kg driven by a 20 b.h.p. V-twin engine which came from an Indian motor cycle. Two examples were built but with no more success. Willys imitated Chevrolet in designing the Willys Air Cooled which borrowed its motor cycle engine from a Harley-Davidson.

Kaiser also tried to take up the challenge with the Midjet Jeep and a slightly heavier model, the Type 1160. Two Midjet Jeeps and four 1160s were built but they did not solve the problem.

One of the reasons for the failure of this programme lay in the use of air-cooled engines generally from motor cycles, with insufficient power, in order to keep down the weight.

For this reason Willys designed and built several lightened models at their own expense on the basis of a standard Jeep. Known by the name MB-L (For Light) Special and nicknamed Gipsy Rose Lee, these vehicles retained the GO Devil water cooled engine and the standard gearbox but the body was reduced to the simplest terms. The tyres were 5,00 × 16 and only the front wheels had brakes. Lightened in this way, the MB-L Special weighed 674 kg.

This time, the efforts made by Willys proved successful but they never received any orders because this class of vehicle had become unnecessary. The test programme and the project for a lightened Jeep came to an end in the closing months of 1943. In a way one can say that Gipsy Rose Lee simply led back to the original Jeep.

The most successful vehicle designed for the use of the airborne troops was achieved in England by Nuffield Mechanisation Limited. It consisted of a Willys MB, the chassis of which had been cut slightly to the rear of the front seats and certain engine equipment modified (battery mounting, air and oil filters, Solex carburetter etc.) The bodywork, while retaining the general lines of that of the Jeep was entirely designed in England. However, the windscreen was that of the American model. To facilitate transport in the confined fuselages of aircraft, the steering column could be removed. Although it performed brilliantly during its tests the Nuffield Willys never went into production.

As part of the investigation into equipment for the jungle war in South East Asia, two British car manufacturers, Standard and SS Cars Ltd. (the future Jaguar Company) built small cross-country vehicles

In 1943 Nuffield Mechanization Limited built this lightened version of the Jeep intended for airborne troops.

Obviously this is not a complete account of the special uses of the Jeep, as it had been designed and built to do everything. Hence the Jeep traveled by railway most usually on tracks with one-metre gauges. Anything larger necessitated more fragile transformations. It was lengthened to carry up to 10 people, it was used to pull ploughs or aircraft, the latter even on the aircraft carriers of the US Navy in the Pacific. It was armour plated and given bodywork of every possible sort and kind to protect its occupants from cold and the weather.

For protection the standard Jeep only had a simple folding top, without any tarpaulin sides nor heating, nor defrosting. Nevertheless, a competent mechanic, if offered certain inducements generally of a liquid variety, could fairly easily improvise a heater fed by a by-pass from the engine cooling system for the sole benefit of the front passengers. Other inventors used

(Continued on p. 69)

In England, the ML Aviation Co. tried to make the Jeep fly by giving it a mechanically driven rotor and some fuselage parts from a glider. This is the Australian prototype for a flying Jeep, the Fleep, photographed in November 1943 before the side doors has been fitted.

THE FLYING JEEPS

Thanks to the defeats inflicted on the Japanese forces in the battles of the Coral Sea and the Bismark Sea, Australia was saved from invasion. On the other hand battles still raged in New Guinea in some of the most inhospitable jungle which exists anywhere.

The nature of the country was such that supplies could only get through by air. So in March 1943 the Directorate of Inventions for the Australian Army, which was run by Sir Lawrence J. Hartnett, President of Holden, the Australian General Motors subsidiary, was given the task of finding a way of delivering Jeeps by air to troops fighting in the Papuan territories.

Dropping by parachute was excluded because the clearings where supplies were landed were very small and the droppings were not sufficiently precise, so Sir Lawrence Hartnett began by studying ways of fitting a Jeep with Autogiro blades which would allow it to land with precision under autorotation.

The Australians were strengthened in their conviction that the project was feasible when they heard that the Experimental Establishment of the British Airbone Forces was carrying out tests with the same objective on a Jeep named the Rotabuggy. However, neither of the two teams could be kept in touch with the work done by the other and they had to work quite independently. John L. Watkins was put in charge of the Australian programme which was called Skywards while the British experiments were conducted by the M.L. Aviation Company. Watkins first thought of using the rotor of a Cierva Autogiro but because of its instability at speeds above 160 km/h he devised new blades with a profile which would support the vehicle in relation to the Dakota which would have to tow it. A single Jeep, serial No. 128276, was allocated to the programme. It was fitted with an aerodynamic structure in wood, a rotor mounting, a clutch and brakes for the three bladed rotor. This assembly could be jettisoned as soon as the machine landed.

The Fleep, contraction of Flying Jeep, was never tested because in December 1943, just as it was ready to fly, the project was abandoned in view of the advances made by the Allied forces in New Guinea.

Similar tests carried out in England had shown that the Jeep could do everything including fly but without any better result in respect of military use. In fact the only Jeep that really flew was the Rotabuggy. The event took place on November 16 1943, at Sherburn-in-Elmet, the towing aircraft being a twin-engined Whitley bomber. Design work had been carried out by Raoul Hafnir.

To find out just what gravity loadings the Jeep could stand, it had been dropped with various loads from a height of 2.30 m. In this way it was found that the Jeep could withstand a force of 11 G without damage, which must be the oddest test ever for its strength of construction.

The rotor of the Rotabuggy had only two blades. The rear of the vehicle was covered by a streamlined fuselage attached to the structure supporting the rotor. This was modified after the first flights revealed serious instability in the Jeep's flying characteristics. After various problems had been solved, the Rotabuggy proved quite satisfactory, but the arrival of gliders capable of carrying vehicles and their crews put an end to the programme.

the exhaust silencer which, ingeniously modified could also furnish its quota of calories. Some of the most ingenious practitioners managed to devise defrosters for the windscreen.

As for the 'coachwork' it was as numerous and varied as the imagination of the constructors and the local resources permitted, without ever offering much in the way of comfort, if one can rely on the memories of drivers and passengers, particularly during the winter campaigns in Italy, Belgium or Alsace. One of the most original bodies was constructed with the aid of parts from the cockpit of a B.17 bomber. The British Army, more conscious of the need to fight the cold than the American Army which does not seem to have made much effort in this respect, got Humber to devise a system of side curtains for the Jeep with hinged doors in tarpaulin and mica.

In 1943 the standard Motor Co. designed this vehicle, a real British Jeep with a 44 b.h.p. engine, but as its front wheels were not driven it had no future, although it performed well.

This photo illustrates better than any words the need to protect driver and passengers against the weather. This is why the Jeep was sometimes called the "Pneumonia Wagon".

Leather helmets and sheep skin jackets were supplemented by a collective poncho in this experimental American Army weather protection which fortunately came to nothing. It is difficult to imagine the crew getting out quickly to start fighting.

Like many war-time Jeeps, that of General De Monsabert had extended mud guards in front. The red, white and blue plate carried the General's stars.

At the Rhine crossing in 1943 this Jeep of the French First Army had wooden sides to protect its occupants. Visibility seems to be at a minimum.

When he commanded the 2nd Corps of the French First Army in Alsace, General Anne-Jean de Goislard de Monsabert had a Jeep with closed bodywork consisting of a metal frame supporting glass windows and roof.

An original conversion made by a French Jeep driver in Germany in 1945 to protect himself from the wind.

A Jeep with a rigid wooden body waiting for American soldiers outside a dentist's surgery in Berlin in 1945.

Bodies like this were made in small batches in the Ordnance workshops of the American Army in Germany.

Members of the 644th Ordnance Company of the US Army built themselves this confortable Jeep with salvaged aircraft parts. It even had a defroster.

In 1947 the American Command in Germany organised a coachwork competition on Jeeps for soldiers who had little to occupy their time. Only salvaged components could be used. Ariel M. Hunt of the 344th Ordnance Company was first of some 50 entries.

Second prize in the 1947 Jeep coachwork competition went to this creation far removed from a standard Jeep.

A Jeep train on the railway between Myitkyina and Baugang in Northern Burma.

A British company of railway engineers converted this Jeep seen in use between Cherbourg and Caen in August 1944. The railway wheels are attached to the standard rims, to increase the track. Without this adaptation the Jeep could only travel on lines of one-metre gauge.

At Luzon in the Philippines, the US Army organised a rail service on the 30 km of track between Bayagang and San Carlos, using what was left of the old rolling stock. This Jeep seen in 1945 could haul trains of 250 tonnes.

Uses for the Jeep Engine

Besides its automotive uses (Allis Chalmers, Marmon- Herrington etc.) the Jeep engine and some of the mechanical units were used to drive electricity generators, welding plant, starter equipment etc.

One of the least known uses of the Jeep engine was its employment after the war to drive a light trolley for permanent way inspection. This trolley which was called Car, Railway Maintenance 60 H.P. Gasoline Engine, by the US Army in February 1946 could carry eight men. The track was adjustable to allow it to operate on seven different railway gauges. Although the technical handbook specified Willys engines exclusively, all the parts lists gave part numbers for Willys and Ford spares so as to use existing stocks of interchangeable parts.

At Mytho amid the rice fields of the Mekong delta in Indo-China the French Service des Materiels et des Batiments Coloniaux used Jeep mechanical parts when building wooden transport vessels to replace the motor junks of the river which were used in great numbers by the units stationed along the creeks where no American or British landing craft were available. Two types of flat bottomed boat were designed to permit navigation in shallow water and landing on banks covered with silt. They were the idea of Captain Hery, carried out by Lieutenant Jeanty. The first was six metres long with an outboard motor of 9 b.h.p. giving a speed of 4 to 6 knots with 1,300 kg of cargo. The second was 8 m long with a Lutetia or Evinrude outboard engine of 22 b.h.p.

60 of these landing craft called the Mytho were built in three months. The hulls were a success but the outboard engines created a lot of problems because they were fragile and there was a lack of spare parts.

"General Delaye, Director of Ordnance for the Land Forces in French Indochina, then asked us to equip the eight-metre boat with a Jeep engine which offered the advantage of instantly available exchange units. The only difficult problem to resolve was that of the transmission, the propeller and angle drive, but once the parts had been designed and a prototype built, quantity production could be carried out by the general workshops of the army. Lieutenant Jeanty spent ten days and part of the nights with his colonial team devising the adaptation. The idea of above board transmission which would have offered the maximum load capacity was quickly abandoned although the mechanical assembly would only have taken up a very small space between the next to last bulkhead and the rear of the boat. This layout needed four complicated changes of direction for the power transmission and was difficult to protect. In the absence of a suitable angle drive they ended up using the standard Jeep gearbox and slightly modifying the reduction ratio of the reverse gear pinion to obtain a speed of 1,100 r.p.m. which was sufficient for slowing down and casting off. Transmission within the hull was arranged by using two differential housings, two shafts and a standard half shaft, providing two right-angle drives. The loss of power was virtually negligible.

The propeller was thus mounted at an optimum height to move the water effectively without wasting power and without sacrificing the benefit of the shallow draught. The engine was sufficiently far back to leave a very useful loading space unobstructed. Two bench seats; which could be removed to accommodate extra cargo, could carry twenty fully equipped men, ten others being able to squat in the middle of the vessel. Tests showed that it was possible to transport 30 men with their equipment or 2.5 tons of cargo at 22 km/h in slack water with a large margin of safety. Thus the Elan prototype was evolved at Mytho and Lieutenant Colonel Malga,

Jack Allingham of the British Ordnance Corps improvised a laundry in the country using his Jeep to work a washing machine which he had found in an abandoned German house.

Willys built this tracked vehicle at the request of the Canadian Army. It was a light unit using Jeep parts, including the engine and some items of the suspension.

Director of Ordnance for the Land Forces in South Vietnam showed it to the General Staff of the Land Forces of the Far East. It emerged victorious from competition with a lighter paddle boat devised by the engineers. The principle of the Mytho boat was adopted. We were invited to build the whole series of 400 landing craft envisaged in several batches over a period of a year but we had to decline the honour because the Mytho workshops still relied too much on hand work. We had to resign ourselves to contributing to the enterprise from a distance by providing several specialists in river boat operations and the chief of our workshop at the 1st Battalion of ordnance repair which, on the recommendation of the Technical Division was given the job of preparing the definitive model and preparing the production line.

The transmission with right-angle drive was abandoned as too difficult to service. The oil in the upper housing had to be watched with extreme care because the slightest deterioration in the seal would have allowed the lubricant to seep by gravity into the lower housing. Moreover, end thrust in the propeller drive shaft could eventually cause play which generated vibrations in the mounting system which fixed the two housings to the hull and could start leaks. A straight drive line was therefore adopted, inclined at an angle which permitted adequate immersion of the propeller. This arrangement needed a stern post and an axial thrust bearing and pushed the engine and gearbox towards the centre of the boat to the disadvantage of carrying capacity. Greater ease of repair had to be paid for by other inconveniences.

From April 1953 the little Mytho boats spread throughout the south and centre of Vietnam » (Extract from Tropiques, the magazine of the colonial troops, January 1955).

The mechanical elements of the Jeep were also used in two projects for tracked vehicles, in Canada. Intended for use by airborne troops, these two compact machines were designed by Willys. The first model, the Tracked Jeep model TJ Mk 1 built in 1943 had its engine mounted transversely at the rear with a Cletrac differential driving the forward track pinions.

These five Mk Is were followed by six Mk IIs, the armour plating and part of the design for which had been carried out by Marmon-Herrington in collaboration with Willys. This time the engine was mounted longitudinally but this vehicle sometimes called the *"Jeep Tank"* got no further than its predecessor. Another Jeep prototype worthy of mention was called the O'Laughlin and was converted into a tracked vehicle with the aid of components obtained from a Weasel.

THE JEEP AT WAR

One of the 25,808 Willys MB of the first series, recognisable by its grille of iron bars, under test by army units in the US.

Jeeps which had crossed the Atlantic in packing cases being assembled in England in 1943.

A unit of the US Coastal Artillery under Captain Steve J. Meade practising the crossing of a ravine by wire at Fort Sheridan Illinois in March 1942. The hard-edged guard under the fuel tank is characteristic of the MB first series.

The same high wire excercise with a first series MB but this time in England. The first series has no Jerrican mounting and the world Willys was stamped at rear (August 1943).

After reassembly on arrival in England the Jeeps received the inscription "Max Speed 40 m.p.h., Caution Left Hand Drive. No Signals."

While awaiting the landings in France the US Army helped the British agricultural war effort by using their GPWs to tow farm machinery.

A Ford converted into a radio vehicle by the British Army.

Practising disembarkation of a Jeep from a mockup of an American Waco glider in preparation for the Normandy landings.

Jeeps, some of them carrying a British airborne trailer on the LST which took them over to the Normandy beaches.

British and American troops practising unloading a Jeep from a Horsa glider before the Normandy landings.

British and American parachute troops reach their objective on the morning of June 6 towing a British Airborne trailer.

This British Jeep landed in Normandy on June 6 1944 has the spade and axe on the bonnet and a Lee Enfield rifle at the side, with the spare wheel in front.

A Jeep ambulance taking German prisoners to the rear on the morning of June 6 1944.

Winston Churchill accompanied by Field Marshal Montgomery in a Jeep on a Normandy beach on June 12 1944.

Field Marshal Montgomery in General Gale's Jeep during an inspection of the Airborne troops, on July 16 1944. The tool clips on the scuttle are typical British army fittings.

Lt. Colonel R.I. Lloyd and the Mayor of Bayeux inaugurating the Bailey bridge across the Aure. The left headlamp of the Jeep has been removed and the right lamp replaced by a black-out lamp, a fairly typical British arrangement.

General de Gaulle finally arrives on liberated French soil on June 14 1944, the anniversary of the capture of Paris by the Germans.

Methodist chaplain Clark J. Wood and his driver George H. MacLain in Normandy, November 1944. The white cross unlike the red of the medical service indicates a military almoner, so does the windscreen's inscription "Chaplain".

American troops on a Jeep crossing the Mayenne on a pontoon operated by the engineers in August 1944.

A Jeep of the "Red Caps", the nick-name given to the British Military Police, in the flooded streets of Bretteville L'Orgueilleuse in Normandy on July 21 1944.

Parisians climbing on to the Jeeps of the liberators in the Rue de Rivoli.

A tank destroyer and a Jeep of the Armoured Regiment of Marine Fusiliers of the 2nd Armoured Division of General Leclerc on the Boulevard Raspail, Paris August 24 1944. The code sign X indicates a vehicle of a General Staff Company.

1-HQ4
28-110-1
1BN-1

The Te Deum at Notre Dame in Paris on August 25 1944 was marred by a burst of gunfire, the source of which remains obscure.

American troops in their Jeeps approaching Paris.

After the triumphal march of General de Gaulle down the Champs Elysees the US 28th Infantry division marches down the famous avenue.

American troops pause for sight-seeing at the Eiffel Tower.

King George VI and Queen Elizabeth visiting Scottish troops on May 21 1944 in a Jeep decorated with flames painted above the silencer.

The Belgian Brigade entering Brussels September 3 1944. Like all Allied vehicles the Jeep has the American star on the bonnet for recognition by Allied aircraft and there is a cockade in Belgian colours on the windscreen.

Equipment of a British communications unit being loaded into an American Waco glider in England on September 16 1944 in preparation for the next day's operation at Arnhem.

Sergeant photographers Lewis and Walker of the British Airborne Division share their rations with a woman from Arnhem on the morning of September 18 1944. Once again the flat bonnet of the Jeep comes in useful.

(Right) British troops pause before entering Eindhoven on September 17 1944.

General Barton commanding the US 4th Infantry Division becomes the first American general to cross the frontier between Belgium and Germany on September 11 1944 during the offensive against Aix-La-Chapelle.

Lt. General Brereton inspecting a Jeep of the 6th Airborne Division which could be parachuted from the bomb bay of a modified Halifax bomber together with a six-pounder anti-tank gun and its crew.

In spite of the war, Christmas must go on. Thanks to the men of the 18th Company of the Medical Corps and the 11th British Armoured Division St Nicolas is able to visit the children of Deurne in Holland.

89
2

The US Army reproduced General Patton's Jeep more or less exactly for the Quarter Master Corps museum at Fort Lee.

General Patton (Left) kept a special Ford Jeep for his personal use with a large red leather seat and two long brass horns on the bonnet.

In February 1944 the Royal Corps of Signals built these special Jeeps with cable drums to set up telephone lines for the British Army.

The parachute harness used for dropping a Jeep, shown at an exhibition of US Army equipment ont the Champs de Mars under the Eiffel Tower in 1945.

General O.H. Lee uses his Jeep's wire cutter to sever the ceremonial ribbon at the inauguration of the temporary bridge over the Albert Canal on February 17 1945.

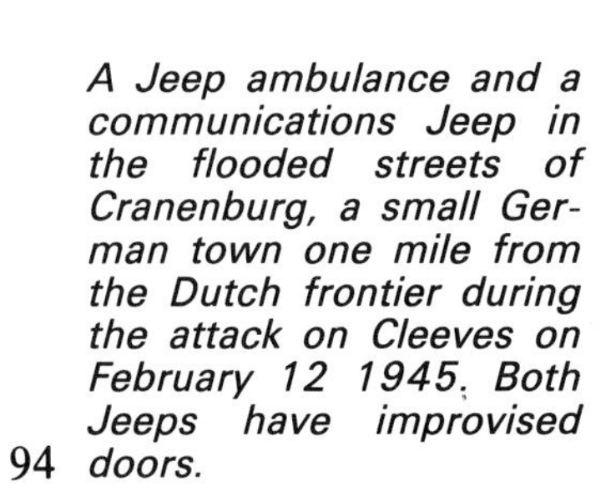

A Jeep ambulance and a communications Jeep in the flooded streets of Cranenburg, a small German town one mile from the Dutch frontier during the attack on Cleeves on February 12 1945. Both Jeeps have improvised doors.

Loading a British Jeep ambulance into a Horsa glider before the Airborne operation against Wesel between Cleeves and Duisbourg on March 22 1945. The stretcher racks are folded down at the rear.

General Horrocks with officers of the 51st Highland Division round a radio Jeep. Note the doors and the special top to protect the radio. (Rees, March 26 1945.)

A Jeep of the US 1st Army crossing the frontier between Belgium and Germany.

German prisoners of war helping to push a Jeep of the British 11th Armoured Division Military Police out of a quagmire.

One of the eight pre-production Bantams fitted with four-wheel steering.

The Ford GP, called the Pygmy or Blitz Buggy. British paratroopers cut the steering wheel like this to fit the vehicle into their aircraft.

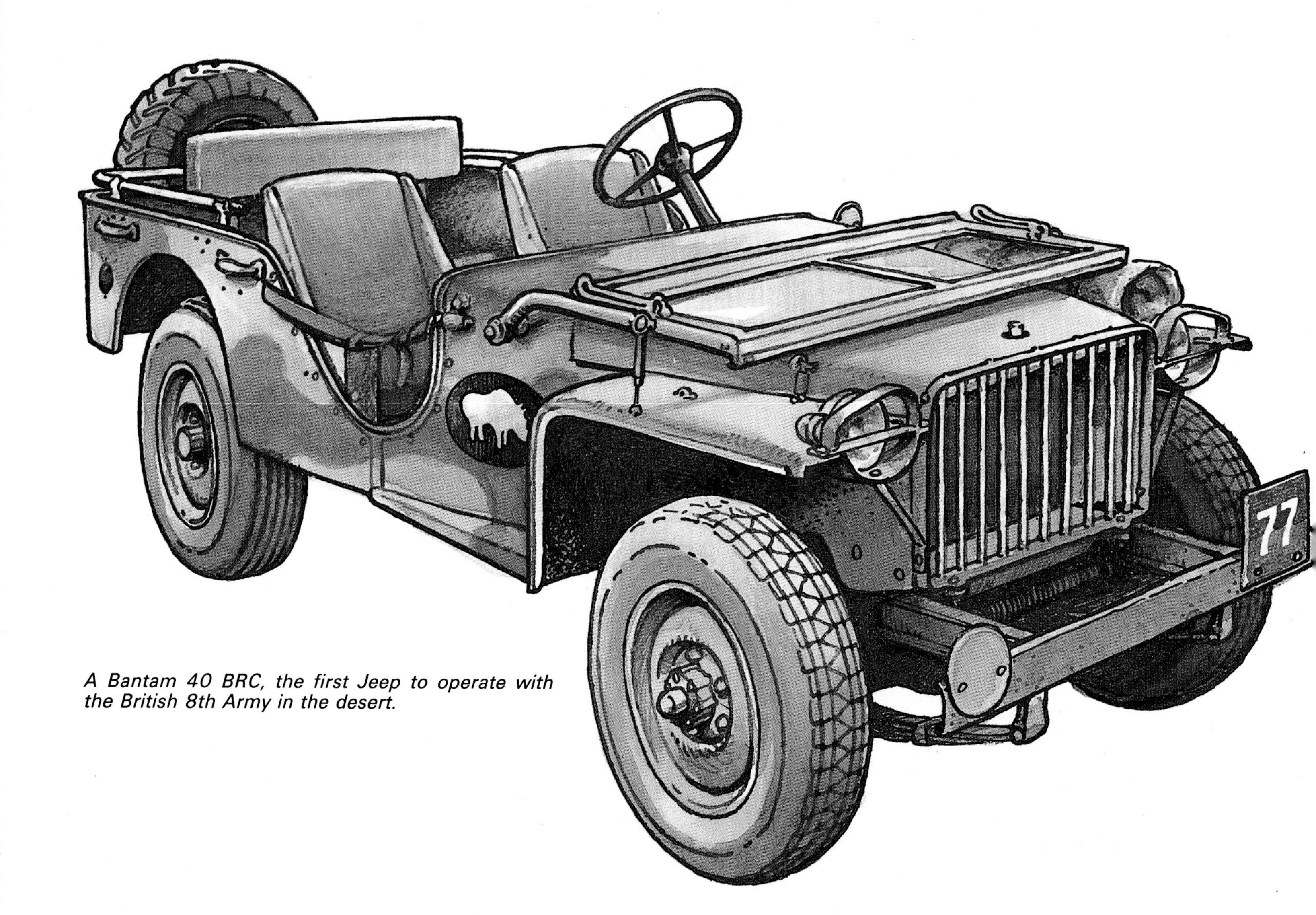

A Bantam 40 BRC, the first Jeep to operate with the British 8th Army in the desert.

Willys MA.

Chaplain
3A 1H
GP 9

2048209

2048209

The M 201 with an armoured squadron ot the Ivory Coast Army in 1963.

The flat bonnet of the Jeep proved very useful as a map table.

Senior officers always found the Jeep a fast and flexible form of transport.

By carrying the camouflage netting on the folding top it could quickly be spread to hide the whole vehicle.

Advance command post of the French 8th Division in 1976.

M 201 with mounting for RASURA radar for battlefield surveillance.

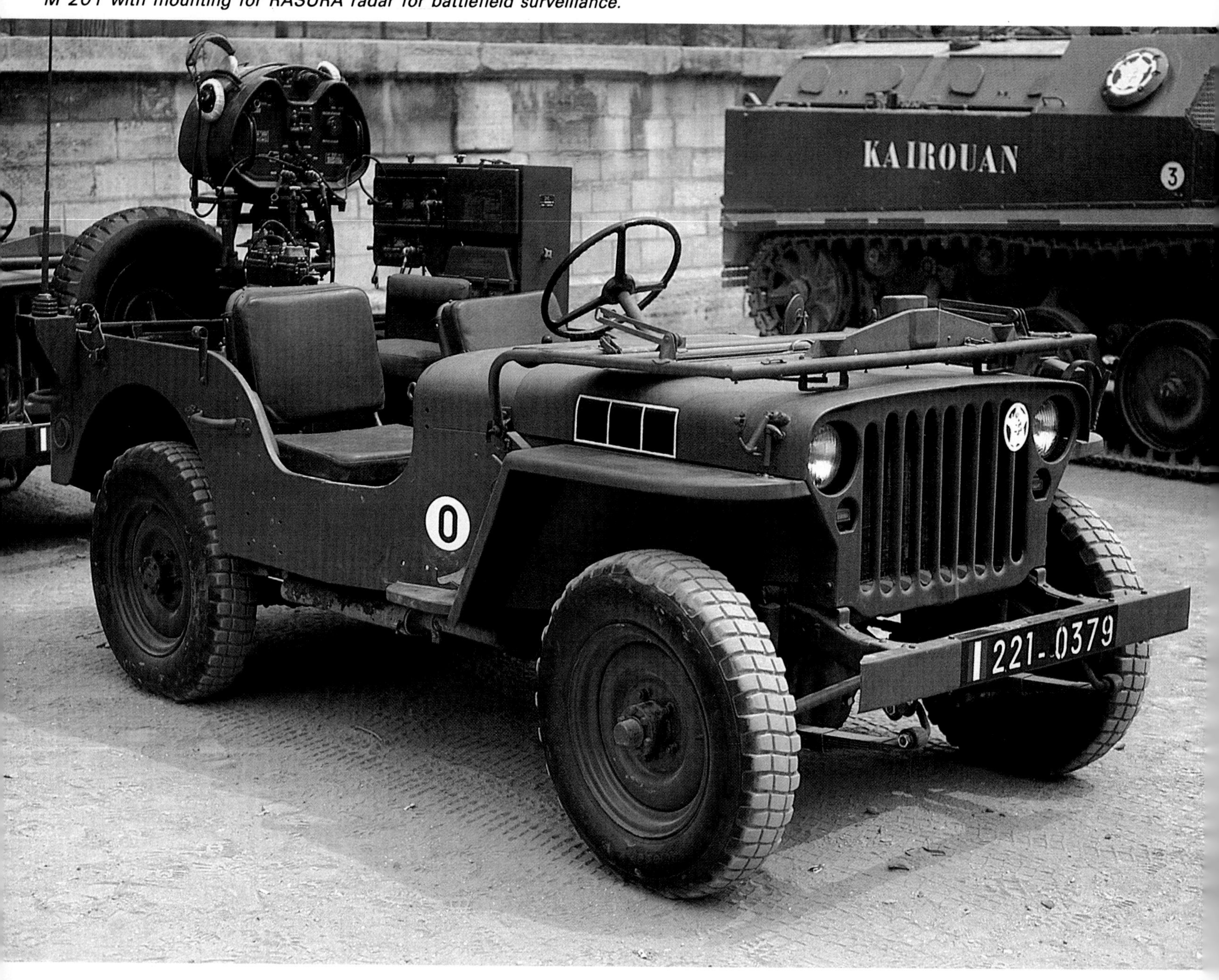

Jeep with launcher for Milan anti-tank missile at Satory 1979.

201 of the 2nd French Armoured Division ith infra-red equipment.

French Army Jeeps in the French Republic of Djibouti were camouflaged in two colours to suit the local terrain.

Jeep of the British 1st Special Air Service Regiment in Belgium.

(Left) Jeep with Entac missile launcher and a missile carrier mounted in front.

De-contaminating an M 201 in training.

M 201 of the 1st Overseas Inter-Services Regiment during mixed Franco-Senegalese maneuvers.

2nd parachute regiment of the French Army in Togo 1967. Instead of combat helmets they wear berets of armoured units.

M 201 with 106 SR of the Chad Infantry Regiment.

ep of the Bataillon
nterie du Pacifique
5 SR in the Pacific.

Armoured Jeep visiting an Algerian village in 1960.

Some US Army reconnaissance units improvised armour plating for their Jeeps especially in the Ardennes campaign.

Like all the United Nations vehicles the M 201 Jeeps of the French contingent in Liban in 1978 were painted white.

The anchor and the tri-colour roundel were used rather than the French flag for sometime to indentify vehicles belonging to the « Infanterie de Marine »

At the end of the war and during the following years US Navy vehicles were painted pearl grey even when appearing in a film like 'Don't Go Near the Water' in 1957.

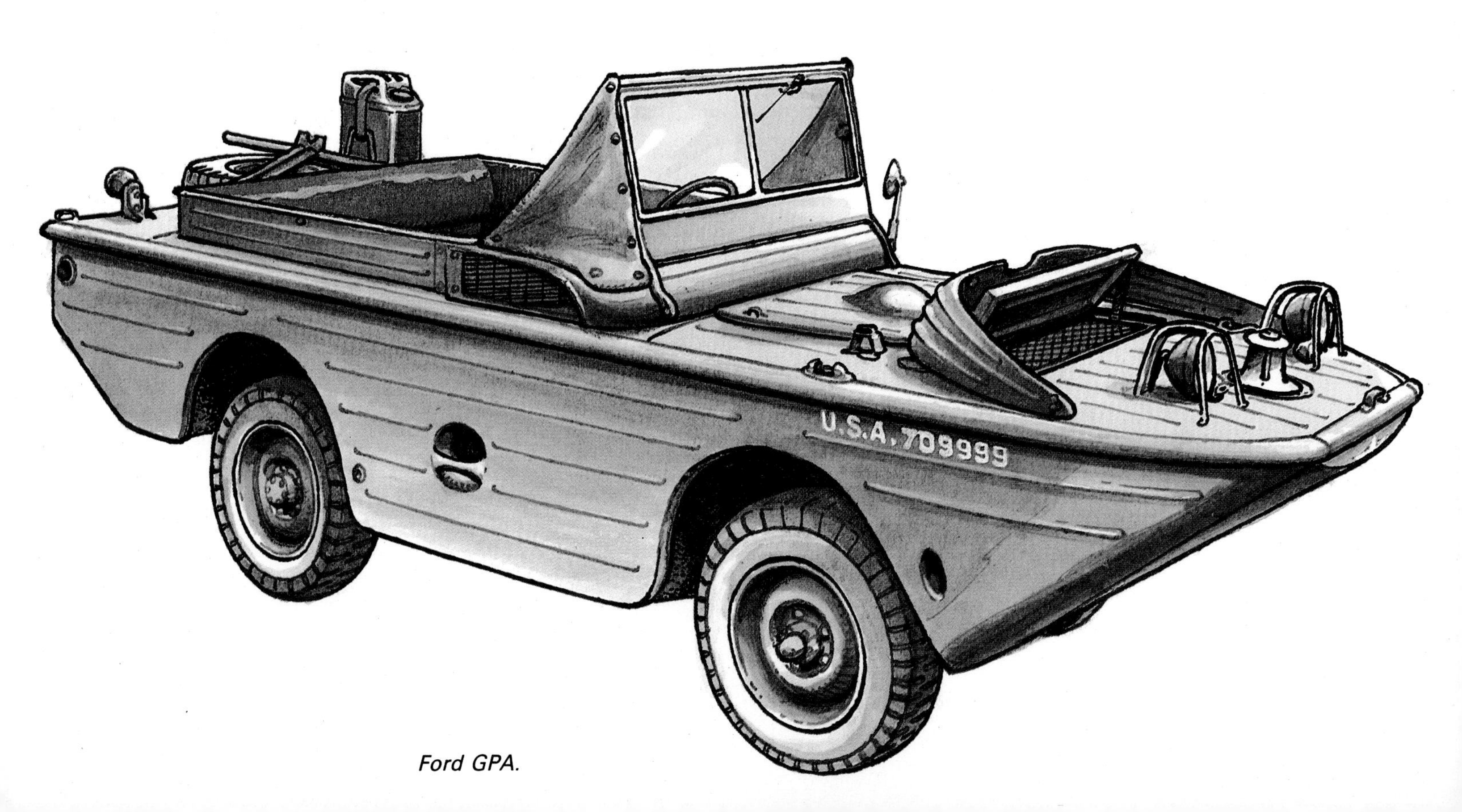

Ford GPA.

GPA used as a flood rescue vehicle by the Maine-et-Loire fire brigade.

Willys MB with CO_2 Fire Fighting Equipment at Paris airport.

Light tanker for forest fires. The reinforced bumper was an accessory designed by Hotchkiss for the HWL Jeeps.

Many of the MB and GPW Jeeps were used by French fire brigades as liaison vehicles or for towing rescue equipment, like this inflatable boat.

A light fire engine of the Eure Department fire service in France.

Willys MB converted into Berglöschfahrzeug (Mountain fire engine) with Rosenbauer pump in Austria 1979.

This Jeep with a flat rear platform was used as an aircraft tractor by the helicopter unit of the French Gendarmerie.

M 201 of the 7th Training Unit of Civil Defense at Brignolles leading a column before an onslaught of a forest fire by the second fire fighting company.

The ubiquitous Jeep serves as transport for a lead crew of a 97th Bomb Group B-17G et Amendola, Italy. Baseball caps worn by gunners helped shade eyes from sun glare.

In 1956 the Jeep starred in the film 'The Tea House of the August Moon' which gave plenty of scope to demonstrate its capabilities.

Journey's end with a Jeep.

Mao Tse-Tung reviewing a real Chinese unit equipped with Japanese tanks in a Jeep abandoned by the Nationalist army.

A Jeep converted into a light truck in England by Farmcraft in 1948.

Finishing line in the Rallye des Cimes.

G. de Bussy in his Jeep with Alfa Romeo engine. ▶

G. DE BUSSY

American patrol with armoured Jeeps near Bastogne, late December 1949.

General John Milikin commanding the US 3rd Corps in the flooded streets of Pont-à-Mousson. Conditions like this led US Army technicians to stipulate 46 cm fording ability without special preparation. The water here is deeper and the driver is trying not to drown his engine.

Some Jeeps of the US 7th Army in Alsace were fitted with 12 4.5 in. rocket launchers like those used on landing ships in Normandy. Rockets could be fired every two seconds. There was a folding support on the chassis beside the driver.

A Jeep with cable reel lays a telephone line in a village in Alsace.

Alsace is liberated and a Jeep of the French Army film service arrives to record events.

A first series MB with wheel extensions for use in soft snow. It also has a wire basket at the rear, improvised doors and a wire cutter in front.

A good demonstration of what the Jeep and its trailer could carry at the Remagen bridge March 7 1945.

The Rhine crossing consumed vast quantities of equipment. In March 1945 wrecked Jeeps and Jeep amphibians were piled up at the roadside to leave space for the traffic.

Jeeps of the French 1st Army re-grouping for the Rhine crossing, March 1945.

Rhine crossing on a ferry built by the engineers for Jeeps of the French 1st Army at Spire, March 1945.

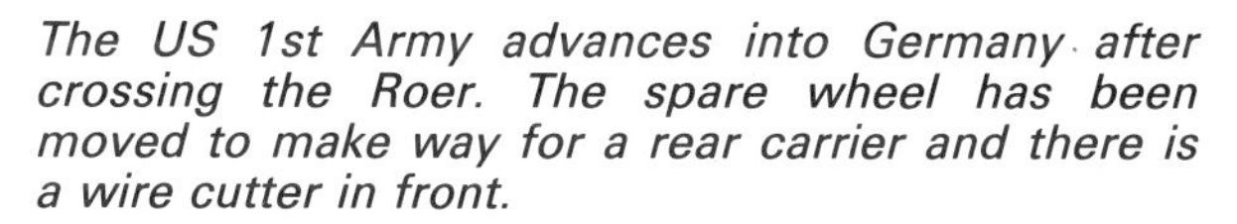

The US 1st Army advances into Germany after crossing the Roer. The spare wheel has been moved to make way for a rear carrier and there is a wire cutter in front.

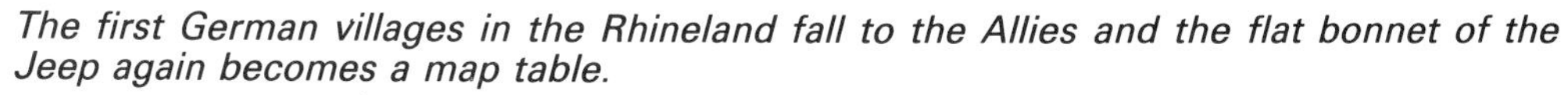

The first German villages in the Rhineland fall to the Allies and the flat bonnet of the Jeep again becomes a map table.

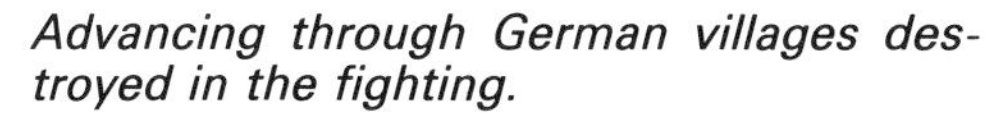

Advancing through German villages destroyed in the fighting.

Inhabitants of a village near Dusseldorf are directed to assemble to receive instructions from the occupying forces by the crew of a Jeep equipped with loudspeaker and amplifier. The feminine figure on the bodywork is a decoration more often seen on American aircraft.

British soldiers distribute drinking water to German refugees from a Jeep in March 1945.

A Jeep captured by the Germans and destroyed soon afterwards by mortar fire.

An exhausted driver falls asleep at the wheel. Beside his seat is the container for a rifle usually mounted under the wind shield. The twin windscreen wipers are worked manually.

Effects of fire from automatic weapons on the tyres of a Jeep. The spare wheel mounting has been moved to the side of the vehicle.

A Jeep used by the press service of the French Army in Germany.

A Jeep passing a column of German transport heading for the sorting depots of the 82nd Parachute Division of the 18th American Airborne Corps on May 3 1945.

Two French Army Jeeps in Austria. The one in the foreground carries a 12.7 mm. machine gun on monopod mounting (May 1945).

Russians and Americans link up at Torgaw in May 1945. The windscreen carries Russian identification marks and they have tool clips on the scuttle.

Before telephone lines have been repaired a service of Jeeps with absolute priority over other traffic carries the mail between Ludwigshafen and the General Staff at Arnhem on April 20 1945.

General Dwight D. Eisenhower signs autographs for American troops who are to be repatriated to the USA some weeks after their liberation from German camps. Note that the driver has shortened the Willys rear mirror mount and fitted it centraly on the windscreen.

American soldiers trying to visit the villa of Adolf Hitler caused serious traffic jams on the road to Berchtesgaden. The Jeeps in the foreground have a variety of equipment in front and on the side to protect the occupants from the mud.

A Moroccan unit of the Armée de l'Atlantique at the gates of La Rochelle soon after it was liberated.

Marines of the Regiment Blindé de Fusiliers-Marins fighting a fire with a hand pump after the La Rochelle pocket was overcome. In the background are some of their armoured Jeeps.

Zouaves under General de Larminat in La Rochelle after its liberation on April 18 1945.

American units enter Royan after a heavy bombardment.

A US Army Jeep carrying American war correspondents in the Royan sector in April 1945.

General Doyan's army prepares to enter Italy via the Val d'Aosta on April 25 1945.

President Roosevelt in a GPW talking to General Patton during the President's visit to Morocco for the Casablanca conference. The ANFA agreements which followed this conference led to the re-equipment of the French army wiht new materiel including Jeeps (January 14-27 1943).

The French forces in North Africa fighting with the Allies in Tunisia had the minimum of equipment, like this Colonel of the 9th Chasseurs still dressed in the uniform of the mechanised troops of 1939 getting into a first series MB in December 1942.

American soldiers in a first series MB in a typical Cairo street scene.

French forces under General de Larminat here receiving a report from Captain de Courseulles received new equipment via the British 8th Army with whom they were fighting and thus became the first French troops to use Jeeps. (January 1943).

At last, equipment has arrived from the US and the French Army prepares for action. (February 1943)

This Moroccan soldier has decorated his new MB with a hunting trophy which may help him to identify it more easily if it "disappears". Units equipped with Jeeps often had their vehicles "borrowed" by units which had not yet received any. (1943)

With the front on the Bari sector stabilised, American troops are received by the mayor of a small Italian village. The extension pipes used for deep water fording have not yet been removed from the Jeep. (September 17 1943).

Equipment of the French expeditionary Corps disembarking in Italy in December 1943.

German prisoners are taken to the rear near Cassino in Jeeps of a Moroccan unit of the French Expeditionary Force during March 1944.

Conditions were particularly difficult in the Italian campaign. The Jeep is being towed out of flood waters by a Dodge Command Car.

A column of the French Expeditionary Force at a halt during the March on Rome in May 1944.

Women drivers of a French medical unit during a march past to celebrate the capture of Rome on June 4 1944.

Men of the 36th Texas Infantry Division at Drammont on August 15 1944 after disembarkation on the coast of Provence.

A Jeep seen in December 1944 with radio antenna moved to allow space for extra Jerricans.

A Jeep escort for British Headquarters Staff on the road in Italy during the winter of 1944.

MP
MP
MILITARY POLICE
5A-101

A Ford GPW ambulance on a jungle airstrip somewhere in the Pacific.

In 1944 British and American Military Police worked with the Italian police and Carabinieri to keep order in liberated towns. As a result of these contacts the Italian police used large numbers of Jeeps for the next 15 years.

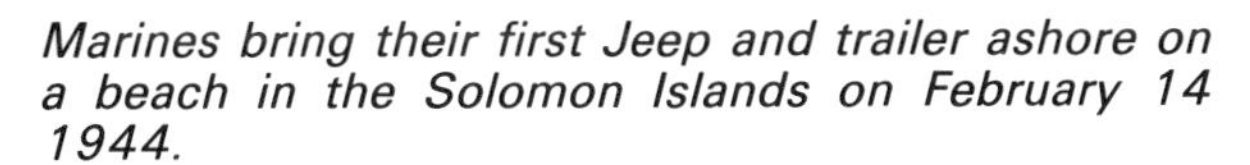

Marines bring their first Jeep and trailer ashore on a beach in the Solomon Islands on February 14 1944.

Mechanics preparing Jeeps for the journey up the Stilwell road opened in January 1945, first road connection between Burma and China for three years.

Lord Mountbatten, the Supreme Commander, Allied Forces, during an inspection in Burma with senior officers. Planks were carried to help in crossing muddy sections.

The Royal Electrical and Mechanical Engineers (R.E.M.E.) used a number of these Jeep-mounted hoists in their workshops. This one is removing cabs from Canadian trucks in Burma.

General MacArthur on Leyte island, October 20 1944. The extension inlet pipe for deep fording operations is folded down in front as drivers rarely had time to dismantle them.

Ormoc on Leyte Island December 14 1944. The island was only taken after a long and costly operation between December 7 and Christmas.

The simplest form of Jeep ambulance; a stretcher attached to the bonnet. British paratroopers at Pegasus Bridge June 7 1944.

JEEP AMBULANCES

Although it was not intended originally, use of the Jeep as a go-anywhere ambulance soon became general. Several conversions were specially designed in the course of the war like the Janes for the British Army, and from the end of 1943 mountings to carry stretchers became a standard fitting on Jeeps assembled in England.

In Australia the Holden Division of General Motors converted 200 Jeeps into ambulances for the campaign in New Guinea.

The various types of mounting can be summarised as follows :

A more advanced casualty conveyance with racks for two stretchers across the bonnet.

- First, the simplest of all, one or two stretchers placed sideways across the engine cover of the Jeep.
- Then, mounted front to rear, either two stretchers in front or two at the rear or permutations of this arrangement.

However, no tarpaulin cover could be erected with these arrangements, so to protect the wounded who were exposed to the open air, the British developed two sorts of cover, one using the standard soft top, the other a top covering the whole vehicle supported by three hoops with the windscreen folded flat (Janes system).

Finally, a set of tubes was provided to erect a rack which could support two stretchers on the upper deck and one down below. The Marines used a special model in the Pacific.

An American Medical Corps Jeep with tubular frame carrying two stretchers lying fore and aft.

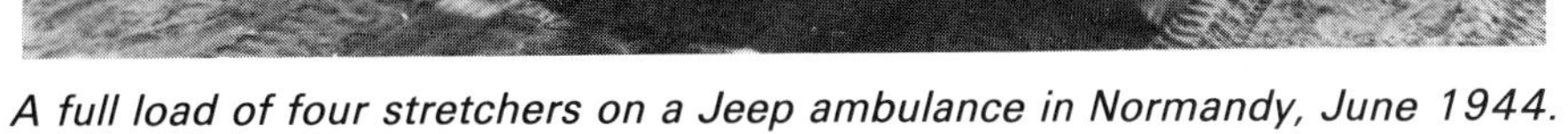
A full load of four stretchers on a Jeep ambulance in Normandy, June 1944.

A simple mounting for two stretchers using two tubes and the hood frame.

The British Army tried to give the wounded a little protection by using the hood and a tarpaulin extension over the stretchers.

A fairly complicated system of folding racks to carry wounded on stretchers.

This tubular framework offered little comfort for the three casualties and made the vehicle rather unstable.

Seen at Duren, Germany, on February 25 1943, this American system at least made it easy to load wounded onto the top deck.

Jeep ambulance in Bougainville, November 1943, with chains on the wheels and a tarpaulin to protect the wounded.

General Rupertus with wounded of the Marines 1st Division Palan Island, September 14 1944. This model, specially developed for the Marines, could carry four casualties, as many as a Dodge ambulance Special racks for stretchers, fixed windscreen, side lockers and spare wheel on the bonnet were features.

This type of Jeep used by the US Army in the Pacific had simple mountings for four stretchers. The plasma flasks held by the orderly were one of the great medical innovations of the Second World War. (Saipan July 1944)

46012

A British SAS Jeep in the Libyan desert. Maximum equipment in minimum space.

THE SAS "WHO DARES WINS"

After joining No. 8 Commando in 1941, David Stirling went to the Middle East to form Brigadier Laycock's *"Layforce"* with two other Commandos. Their main missions were commando-type actions but because of unforeseeable circumstances Layforce suffered setbacks in Crete and in the Western desert. As the Middle East Command could no longer support the requirements of the Layforce it was decided to disband it. Just before this happened, David Stirling was injured in landing after a practice parachute jump and while confined to hospital for several weeks he spent the time writing a report for the British High Command in the Middle East.

His main idea was that most of the objectives for which Layforce was intended, some of them of fundamental strategic importance, could be attained by a unit with no more than a twentieth of the 1,600 men intended for Layforce. His plan envisaged commando operations simultaneously on the three airfields at Gazala and the two at Tmimi where almost all the German combat forces were concentrated. These operations were to take place simultaneously during the night preceding the offensive planned by the Eighth Army. The operation as planned by Stirling required only 65 men, five bombers to drop them by parachute and a patrol of the Long Range Desert Group to evacuate the men at the end of the operation.

This is how the Special Air Service came to be created, with a green light from the British High

The SAS Jeeps had incredible fire power; twin Vickers K guns at the rear, another in front and a 50 calibre aircraft gun.

Armament varied according to the use of the vehicle. This Jeep was a spare fuel carrier for the Commando and had only a twin Vickers K mounting.

To help in deep penetration of territory held by the Germans and Italians the SAS disguised a Jeep as a German Kubelwagen. It looked as if the driver was sitting in the rear.

Coolant expansion tank, part of the US Army Desert Cooling Kit.

Command. This first mission, which took place on a moonless night with a 90 km/h wind and a sandstorm was a real fiasco but during the ensuing weeks David Stirling, with the support of Major General Reid and of the Long Range Desert Group, successfully carried out a whole series of raids on the German and Italian airfields on the coast between Agedabia and Tamet.

The method at that time was simple. A patrol of the LRDG dropped a group of four men at 15 to 20 km from the aerodrome chosen as a target. Having reconnoitered the target, they inserted into the fuel tanks of the aircraft small Lewes plastic incendiary bombs fitted with detonators timed to go off 2 1/2 hours later. Later the interval before the explosion was reduced progressively to half an hour, so that all the aircraft blew up at about the same time but still left sufficient time for the SAS to withdraw.

Having demonstrated the effectiveness of these SAS raids, the Stirling Commando was joined by a detachment of French parachutists at the end of 1941. They had been trained in England and were commanded by Georges Berge. Stirling wrote about them *"this small group of men had an enormous influence on the operations of the SAS and became an elite element in the SAS regiment. Their intuition, the originality of their thinking and their courage supplied basic elements in the concept of the SAS. I well remember they day when I went to Beirut to obtain the consent of General de Gaulle for the enrolment of the Free French in the SAS. Before giving his permission he asked me many questions about the role of the unit and about our operational methods."*

In 1942 Stirling achieved independence in relation to the equipment for his 130 men. So the SAS became independent of the LRDG. The equipment consisted of 20 3-ton Bedford trucks for transport and 16 Jeeps which were equipped with a twin set of a Vickers K and a Browning 50 aircraft gun taken from a wrecked aircraft. This armament was used for direct attacks on enemy targets and for breaking out in the event of unfortunate encounters during the journey. At the end of the year in spite of a disastrous setback in Libya, Stirling learned that he was promoted to Lieutenant Colonel and that the SAS had been given the status of a regiment and was permitted to recruit more men. David Stirling was also made the commanding officer of the SAS, of the remains of the Middle East Commando and the Greek Sacred Squadron, nearly 800 men in all. But Stirling also had to keep his promise and part with the French squadron.

In the early months of 1943 David Stirling was taken prisoner near Gabes in Tunisia. After a slight uncertainty the regiment undertook several commando-type operations under the command of Col. Mayne before returning to its real function. At about the same time the 2nd SAS Regiment which had been formed in England by Bill Stirling arrived in Tunisia to carry out strategic raids on Sicily and the Italian coast.

Sometime afterwards the 1st and 2nd SAS Regiments were repatriated to Great Britain to be regrouped and doubled in size. Two battalians of Free French, a Belgian group and the GAQ regiment who provided the indispensible communications joined the ranks. This raised the SAS to brigade strength.

After various discussions, Stirling's philosophy was adopted and in Europe the force was utilised on a more ambitious basis.

The Greek "Sacred Battalion" fighting with the Allies distinguished itself on commando raids with these special Jeeps.

The French also engaged in successful SAS-type missions in Belgium and against the Atlantic coast pockets as here at the liberation of La Rochelle.

First, small reconnaissance groups equipped with radio were dropped by parachute. If their reports were favourable the group was immediately reinforced, first by more men and then if the ground was suitable, armed Jeeps were dropped. With their fire power and their very low profile, the Jeeps were a formidable asset and could inflict great damage on the enemy to say nothing of the possibilities they offered in enlarging the sphere of operations. The bases were supplied with great efficiency by the RAF.

Once the bases were set up (there were 43 of them in France), small groups, generally squads of four men as described by Stirling and mounted on a Jeep were sent to harass enemy communications wherever they could find them; roads were mined, railways lines blown up and transport convoys ambushed. Targets for bombing were communicated to the RAF. Finally it was the job of the SAS to arm and train French Resistance groups. In a letter which he sent to Brigadier MacLeod, General Eisenhower, Supreme Commander of the Allied Forces wrote *"the ferocity with which the enemy has attacked the troops of the SAS has shown clearly the extent of the damage that you have been able to inflict on the German troops both by direct action and by the information that you have supplied on the positions and the movements of the German armed forces."*

Two Jeeps were also parachuted into France by the American OSS for the use of their units working with the French Maquis.

One of the four prototypes for an amphibian Jeep developed from the experimental Marmon-Herrington QMC-4.

THE AMPHIBIAN JEEP

Apart from the theoretical and practical problems reviewed in Chapter One, the mechanisation of reconnaissance units increased the difficulty of crossing stretches of water because whereas a cavalry patrol could usually cross a stream or a river fairly easily, provided the current was not too strong, this presented an almost impossible obstacle for a mechanised unit. So the idea of an amphibian vehicle was taken up by many designers who tried to resolve the contradictory problems posed by this type of vehicle.

In fact the shapes of hull which are the most efficient for navigation are generally the least suitable for travel across country. The layout of the mechanical parts requires holes to be pierced in the hull at many points which pose a whole series of problems, especially in regard to watertightness and corrosion. Maneuverability and stability in the water depend upon features which became serious handicaps on dry land.

Every amphibian vehicle is more bulky than the land vehicle from which it is derived, because of the necessary and obvious application of the theorem of Archimedes to confer buoyancy on an object which is far from having any in its original state.

Finally it is expensive because it requires special means of propulsion, a much higher standard of wordmanship in the welding and assembly of the hull and a superior quality of metals and other materials in its construction, especially on a machine destined for navigation in the sea or in brackish water.

An amphibian Jeep in its element with breakwater in position and triangular fabric curtains protecting the passengers. The open air intake is only permissible in calm water. Spray indicates that the wheels are being driven for additional propulsion.

Members of the SS testing a Trippel SG 4 developed for their use with Opel Kapitan mechanical parts on the lake in Grunenwald, near Berlin, October 24 1936. It is carrying seven but still has a good freeboard.

The first experiments in this field date from the years immediately preceding the first World War. The boat-car of Ravaillier tried out in 1910 on the lakes at Versailles and on the river Rhone demonstrated the feasibility of the idea, but the vehicle was never put into production because potential users considered its payload insufficient.

As we shall see, this factor limited the use and therefore the prospects for amphibian vehicles until the end of the forties. If one excludes combat vehicles, it seems that it was in France that the first plan for an amphibian reconnaissance vehicle was considered by the military. This plan, drawn up by the Directorate of Cavalry in June 1933, produced a curious vehicle built by an inventor at least as quaint as his invention, Monsieur Texier de la Caillerie. Although it worked satisfactorily during its trials in 1935 the vehicle was not adopted, as its type of construction, its interior layout and its lack of carrying capacity ruled out any military use.

So, on the eve of the second World War we knew how to build light amphibian vehicles, how to operate them on land and in water, provided the current was not too strong and the slope for climbing out of the water was not too steep, but none of these vehicles was suitable for normal military service. The sole exception, the Trippel amphibian car, was then being tested in Germany, but although it represented a clear advance on the machines then existing, the 1939 Trippel SG 4 was still far from being a suitable vehicle to put into the hands of all and sundry for war service.

It was only after a design study commissioned from Porsche by the Heereswaffenamt (Ordnance Service of the German Army) in 1940 that a satisfactory — indeed exceptional — amphibian vehicle emerged in the shape of the Schwimmwagen. After 150 pre-production models (Type 128) had been built, the final model (Type 166) was delivered to army units starting in 1941. 14,625 examples were delivered, a production figure of the same order as that for the amphibian Jeep.

In the United States, the question was approached on parallel lines with similar results Credit for being the first to grasp all the essentials of the problem must go to Robert W. Hofheins of Buffalo in the State of New York who worked on it from 1940 onwards. His efforts came to nothing in terms of series production and his company, the Amphibian Car Corporation had to close down, although several satisfactory prototypes were built.

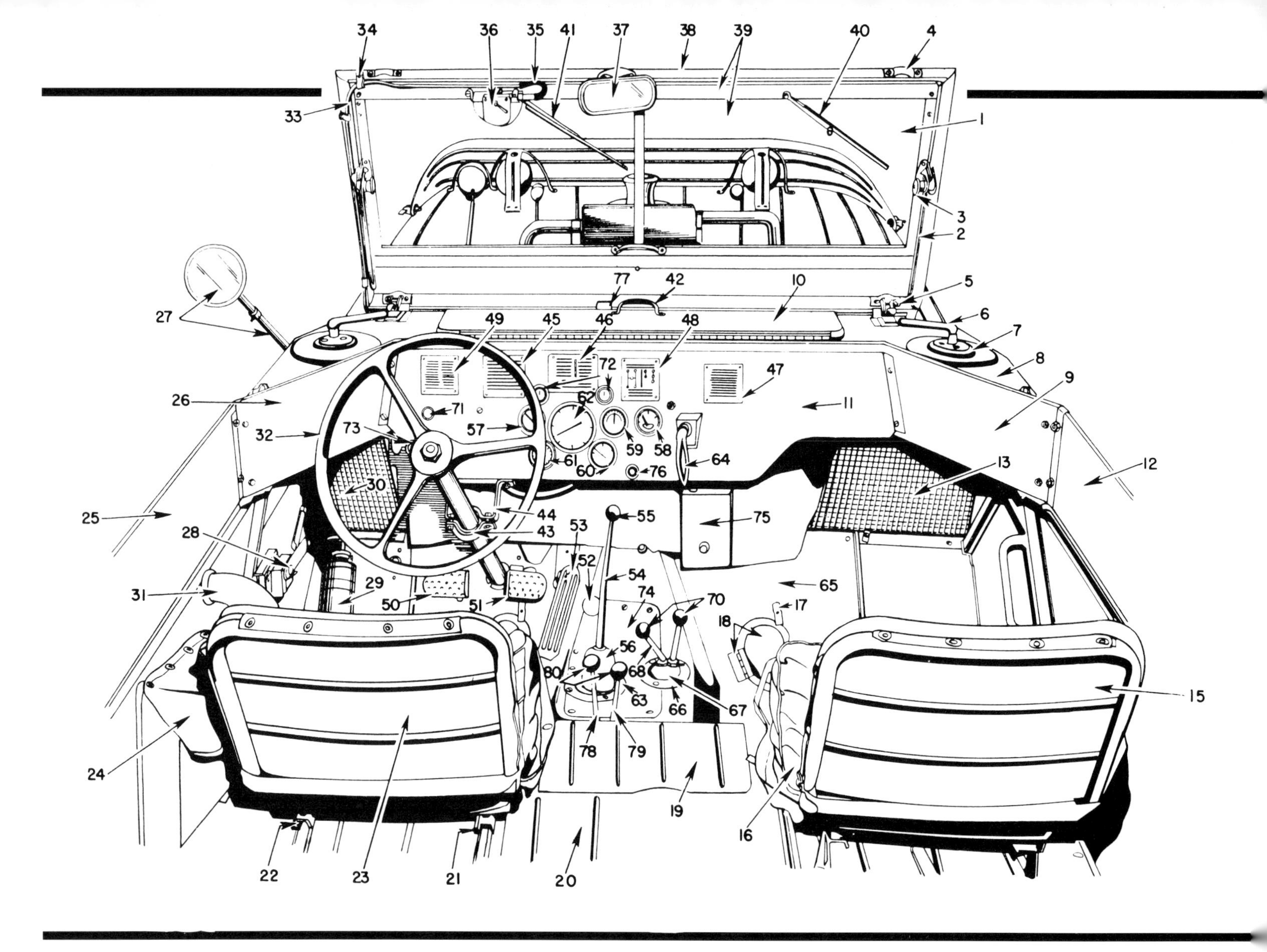

The Design of the Amphibian Jeep

As early as 1940 it was accepted in theory that an amphibian version of the Jeep was worth studying but it was not until 1941 that financial credits could be obtained to get the project started.

The NRDC was made responsible for examining the idea. The National Defence Research Committee was an organisation composed of research workers and civil engineers charged with finding practical solutions to many problems raised by the industrial mobilisation of the United States. The NRDC's ability to innovate was guaranteed by the origins of its members, who had been recruited from private industry and the universities, and by the absence of any ties subordinating it to military technical organisations or the American civil service.

Responsibility for the design was entrusted to P.C. Putman of the NRDC who enlisted the help of Roderic Stephens Jnr. one of the proprietors of the Sparkman and Stephens boatyards who specialised in the design of hulls for racing yachts. This team worked later on the DUKW project. The design of

1 — Windscreen.
2 — Support strut for windscreen.
3 — Clamp screw for windscreen strut.
4 — Clamp for folded windscreen.
5 — Clamp for erected windscreen.
6 — Rotary control for side ventilation system.
7 — Cap for side ventilation system.
8 — Upper deck.
9 — Front face of driving compartment.
10 — Trap over cooling air outlet for amphibian use.
11 — Instrument panel.
12 — Side face of driving compartment.
13 — Right engine air intake for amphibian use.
14 — Number not allocated.
15 — Passenger seats.
16 — Inflated air cushion.
17 — Catch securing inspection plate.
18 — Inspection plate.
19 — Inspection plate for pump and gearbox compartment.
20 — Rear deck.
21 — Right Adjustment runner for driving seat.
22 — Left adjustment runner for driving seat.
23 — Driving seat adjustable fore and aft.
24 — Canvas container for hood canvas.
25 — Left face of driving compartment.
26 — Left front of driving compartment.
27 — Exterior driving mirror.
28 — Control lever for main engine air intake.
29 — Fire extinguisher.
30 — Left engine air intake for amphibian use.
31 — Bilge pump outlet pipe.
32 — Steering wheel.
33 — Metal pipe for suction screen wiper.
34 — Securing clips for suction pipe.
35 — Rubber suction pipe for windscreen wiper.
36 — Vacuum wiper motor.
37 — Interior driving mirror.
38 — Windscreen frame.
39 — Glass and inner frame of windscreen.
40 — Hand wiper for windscreen.
41 — Windscreen wiper blade.
42 — Handle for air exit flap.
43 — Steering column support clamp.
44 — Steering column support.
45 — Plate showing maximum speeds in gears.
46 — Plate with instructions for use of propeller and pump.
47 — Plate with instructions regarding voltage, no regulator being fitted.
48 — Plate with instructions on use of the gearboxes.
49 — Identification plate.
50 — Clutch pedal.
51 — Brake pedal.
52 — Foot rest at accelerator position corresponding to maximum permitted speed (about 60 km/h).
53 — Accelerator pedal.
54 — Gear lever.
55 — Gear lever knob.
56 — Rubber gaiter for gear lever ball.
57 — Fuel gauge.
58 — Volmeter.
59 — Ammeter.
60 — Thermometer.
61 — Oil pressure gauge.
62 — Speedometer.
63 — Securing ring for rubber gaiter over gear lever ball.
64 — Handbrake.
65 — Foot board.
66 — Securing ring for leather gaiter on transfer gear lever.
67 — Leather gaiter on transfer gear lever.
68 — Lever engaging drive to front axle.
69 — Lever engaging low range in transfer box.
70 — Knobs on transfer box controls.
71 — Manual starter.
72 — Black-out lighting for instrument panel.
73 — Switch for parking lights, headlamps and black-out light.
74 — Gearbox inspection plate.
75 — Filter.
76 — Clutch control for mechanical capstan.
77 — Return spring for ventilation flap.
78 — Control lever for propellor.
79 — Control lever for bilge pump.
80 — Knobs on propellor and pump controls.

the hull took shape in the New York establishment of Sparkman and Stephens Inc. while work on the adaptation of the mechanical parts of the Jeep was carried out by Marmon-Herrington. The vehicle was given the designation QMC-4 (QMC for Quarter Master Corps, which among other things was still responsible for investigations into motor vehicles). The first tests took place in 1941.

Marmon-Herrington which called itself *"the mechanisation laboratory of the United States Army"* was founded on March 13 1931 when Walter C. Marmon decided to provide financial support for the research work A.W. Herrington was doing on cross country vehicles.

A.W. Herrington had served as a captain in the American Army in France during the first World War, where he had plenty of opportunities to observe the poor performance of the transport vehicles of that time the moment they had to travel on anything but a hard road surface. On his return to the United States, Herrington joined with the arsenal at Holabird in some experiments aimed at improving the performance of the vehicles then used by the U.S. Army, such as the Liberty truck, the Army being conscious

A Schwimmwagen used as a liaison vehicle by a Panther regiment on the Russian front because of its excellent cross country performance. It has big 200 × 16 cross country tyres instead of the usual 5.25 × 16. Among the large amount of impedimenta is the famous paddle provided because there was no reverse in water.

both of the lack of mobility of the vehicles it had available and the impossibility of obtaining credits for new vehicles from a profoundly pacifist Congress.

After the foundation of Marmon-Herrington the design and construction of new vehicles was intensified, particularly for the Middle East (Iraq Petroleum Co., Iranian Army) the vehicles being entirely designed by Herrington.

In 1935, Marmon-Herrington took a new step forward by developing an off-road conversion kit adaptable to the chassis of the road going trucks then in quantity production by the big American manufacturers.

Although Dodge and International chassis were converted, it was Ford who provided the main support for these conversions and a cooperation agreement was soon signed by the two companies. This is how Marmon-Herrington came to design the prototype of the amphibian Jeep while ar Ford, arrangements for its construction were being made by F.G. Kerby and C.L. Kramer. The first Ford prototype was delivered to the U.S. Army on February 18 1942.

An amphibian Jeep evacuates casualties during maneuvers in North Africa by the American Army.

The same vehicle alongside a medical shelter on a GMC chassis in 1943.

After the ANFA agreements, large quantities of American equipment were supplied to the French forces. Here are two amphibian Jeeps parading before the military authorities in Morocco.

Jeep amphibians being maintained by women members of the British Army. The vehicle in the foreground has the engine compartment open (April 25 1944).

Production of the Amphibian

The production model was called the GPA. (General Purpose Amphibian). The first contract for 7,896 units was placed on April 10 1942 (contract No. 398-QM-12 937. U.S. Army Registration Nos. 702,104 to 709,999). The next contract W-374-ORD-2782, was for the construction of 2,104 GPAs which were given American army registrations 7,010,000 to 7,012,103. These first orders for 10,000 vehicles were followed by another for 2,785 according to Ford or for 2,788 according to the U.S. Army registration numbers allocated, taking the total number of GPAs ordered up to 12,785 or 12,788 according to which source one accepts. Most of the vehicles were sent to foreign theatres of operation. The first vehicles left the line in September 1942 and production ceased in June 1943.

A British Army convoy en route for the south coast to embark for the Continent. The Jeep amphibian carries the number of the landing craft on which it will travel and the rope in regulation position attached to the bollard on the prow which was the sole means of towing a Jeep in the water.

Use of the Amphibian Jeep

Like all amphibians, the GPA was more difficult and more expensive to make than the equivalent land vehicle; it was 606 kg heavier with practically the same engine, it had very little freeboard, it could only carry a small useful load and unlike that of the land Jeep, its body did not lend itself to conversion. Finally, the need for an amphibian reconnaissance vehicle had dwindled.

It had been hoped that the GPA would provide reconnaissance units with a universal vehicle which would allow them to keep going in difficult conditions and avoid hold-ups due to breaks in the road system caused by war operations or natural forces. Unfortunately it was soon found that although it is quite easy to put an amphibian into the water and even to navigate once it is there, getting it out of the water can be a very different matter. In the United States and in Russia, many rivers have gently sloping banks but it is not at all the same in Western Europe where most rivers have embankments and where big rivers have been confined between vertical banks like canals.

The amphibious vehicle thus turned out to have very little use in European operations except on the Russian front where a large proportion of the GPAs were sent under Lend Lease. They were actually copied by the USSR. In any case the small carrying capacity of the GPA meant that practically its only useful role was as a liaison vehicle during landing operations.

A US Army Jeep amphibian near the Eiffel Tower on August 25 1944.

Russian soldiers push an amphibian Jeep stuck on a muddy river bank. The slight slope contrasts with river banks in Western Europe.

An amphibian Jeep of the Israeli navy in Haifa for the march past on April 19 1949. These war surplus vehicles had no windscreen or breakwater and were called the Goose in Israel.

Technical Description

It was a tricky problem fitting everything into the steel hull because of its small size and hence the limited space available for the mechanical elements but in spite of it all, the space in the passenger compartment was practically the same as on the land Jeep.

Propulsion in the water was by a propeller mounted on a shaft emerging from a tunnel in the rear of the vehicle. The shaft was driven from the transfer gearbox. Rotation speeds thus varied with the gear ratio selected by the driver. This method of construction had the advantage of giving the vehicle a reverse gear in the water. A certain amount of additional propulsion effort could be obtained by driving the wheels in the water but if necessary these could be disconnected to avoid throwing too much water into the load compartment. Use of the front wheels to drive and steer, together with the rudder and propeller, helped in keeping the vehicle at right angles to the bank when climbing out of the water. The driver merely had to concentrate on his driving as he had no special actions to take to convert the Jeep from a motor boat to a land vehicle.

The small freeboard which limited the GPA's aquatic performance was often criticised. To keep the vehicle maneuverable on land the dimensions of the hull, particularly the width, had been limited and did not enclose a sufficient volume to ensure adequate bouyancy. In fact it was the four tyres acting as buoys which provided the essential element in its flotation.

Continued p. 174)

Half Safe in the Champs Élysées in the original form with extended prow in which it crossed the Atlantic.

"HALF SAFE"

It was certainly the craziest trip ever undertaken with a Ford GPA; nothing less than the crossing of the Atlantic. Ben Carlin, an Australian engineer bought a war surplus Ford GPA in 1947. He took two years to restore it to new condition and modify it for the great adventure. The main features were an additional fuel tank in front, in the form of a prow, a water tank forming an extension of the rear deck and a cabin round the seats which he fitted out with the minimum of equipment to sustain life. Deciding that it was best not to undertake this adventure alone, he was joined by an American girl friend named Elinore who had come back from China and accepted his crazy proposition without even having seen the vehicle. After tests and modifications and a search for finance, Ben Carlin and Elinore departed from Halifax on July 19 1950 after telling the Canadian coastguards that they were going fishing.

The first stage, which was the longest, consisting of 1,500 nautical miles was to take them as far as the Azores. For this Ben Carlin had fitted his GPA or rather « Half Safe » because that was the name he had given it, with two additional fuel tanks. The first, holding 350 gallons (2,050 litres) was slung under the GPA to form a keel; the second was an aircraft drop tank which he towed. The first twenty days of the voyage went by without serious problems beyond sea sickness, a noisy tappet, dirty plugs and a blown head gasket.

Every two or three days came the ceremony of filling up the fuel tank from the tank they were towing and a change of oil. The GPA cruised at 3 knots and used 1,6 litres of fuel per nautical mile (86 lit/100 km.) The 3,410 litres of petrol in the tanks were sufficient to reach the Azores. The first problem to arise was one of water. The rear water tank had sprung a leak and from then on Ben and Elinore had no more than 15 litres of drinking water. Then on August 9, while they were still 750 miles from the Azores came a crisis. Ben had to transfer all the fuel in the reserve tank underneath the Jeep into the aircraft tank and the tanks on board. To avoid any change in the centre of gravity which could have been caused by the creation of an air pocket in the keel tank, Ben had planned to replace the fuel by sea water as the transfer proceeded. The long and perilous operation was successfully accomplished. But when he tried to dump the keel tank, and released the anchorages on one side, the tank sank and nearly dragged the GPA with it. The boat was listing at 30 degrees, water was pouring into the cabin, the sea was rough and something had to be done quickly or they would sink. Elinore climbed onto the roof of the cabin to act as a counterweight, while Ben dived overboard with a knife to cut the remaining lashings. The tank sank safely, Ben and Elinore had got away with it and began mopping up.

They arrived at Flores in the Azores on August 19 after 32 days at sea. They had intended to spend two days there but stayed for more than a week. From island to island they continued their voyage across the Azores, then on November 18 they left St. Miguel for Madeira. From the first night their troubles began. The prow tank was stopped up and became unusable. A few days later worse happened. In the middle of the night Ben made a mistake and set adrift the aircraft tank which they were towing. There was no hope of finding it again. If the sea remained calm, Half Safe would have just enough fuel to reach Madeira but it was now December. The sea became rough. All of a sudden the motor coughed and then stopped. Ben opened the engine cover and saw a plume of smoke coming out. The fuel pump was leaking. He repaired it but when he tried to restart there was an explosion. Ben put out the flames with the extinguisher but he could not see anything. Fortunately the Jeep had not caught fire and Helf Safe continued its voyage. The weather became worse and the sea became more and more rough. Radio contact became impossible and already Madeira had announced that they were lost. The nightmare continued for three days and two nights. Ben asked Elinore if she wanted to give up. No. They would keep right on to the end.

Once they were clear of the storm another problem arose; they no longer had enough fuel. Fortunately, they were able to get supplies in the open sea from a Portugese tramp steamer. At last they reached Madeira and soon afterwards, Africa.

Half Safe made contact with African soil on February 22 at Cape Juby. The Atlantic had been conquered. Even so, Half Safe still had problems. The standard springs designed for the 1,6 tons of the GPA had to support more than two tons but on land the problems seemed less serious. Ben and Elinore drove up as far as Gibraltar, then to Lisbon, Madrid, Bordeaux, Paris, Brussels, Hamburg and on to Malmo in Sweden before crossing to London and then up to Birmingham. Everywhere they received a triumphal welcome. They were now on the publicity roundabout and they were invited to cocktail parties, exhibitions and two Concours d'Elegance, at Deauville and Knokke Le Zoute. On arrival at Birmingham which was the end of the trip, Ben Carlin who had learned a lot from this journey, rebuilt Half Safe entirely with the intention of completing a trip round the world. He took two and a half years to get everything ready but his hopes of selling the project to the motor industry came to nothing.

Ready to continue its world tour, Half Safe passes the Eiffel Tower on the Seine.

After having crossed Canada and the Atlantic via the Azores, landing at Portugal and touring Europe, Half Safe was rebuilt in England, did a world tour and appeared again in Paris at the Salon des Sports on April 25 1955.

The fire brigade of Vierzon, Cher used a GPA until 1976. It was in practically standard trim except for a cylindrical rubber seal for the shaft driving the propeller.

For these reasons, the vehicle was provided with powerful bilge pumps (one pump driven by the shaft which drove the propeller and an emergency hand pump). These got rid of the water although the compartments of the GPA hull could never be dried out completely. The main pump could draw water from the engine compartment, the passenger compartment or from both at the same time by operation of a valve. In addition there was a curved wooden shield which could be lowered at the front of the vehicle to act as a breakwater. In course of production this wooden shield was replaced by one in ribbed sheet steel.

The engine was cooled by water radiator (capacity 10.4 litres) and by forced air circulation. Air was drawn in through an intake of considerable size on the front deck of the hull and was evacuated through side grilles level with the front seats. During operations on water, the breakwater diverted part of the airflow away from the intake and the possibility of

This GPA built in 1942 was used by the Fire Department at Jamestown, North Dakota, until August 1975. It was adapted for flood rescue work and played an important part during the floods of 1969. There are steps on the hull and extra seats fore and aft. Like many US fire fighting vehicles it is painted white.

The Maine-et-Loire fire brigade used several GPAs for flood rescue work. This one has exhaust outlets in front and the orifice for the starting handle can be seen by the rear wheel. One of these vehicles was presented to the Armoured Vehicles Museum at Saumur.

taking in water made it necessary to keep it closed most of the time. In this case, air could be taken to the fan from two intakes in the driving compartment, emerging through a grille behind the windscreen.

The body was in sheet steel treated against corrosion and the nuts and bolts were cadmium plated.

In the water the GPA was steered by the normal steering wheel which turned the front wheels and operated the rudder at the rear by means of a cable and pulleys, the rudder naturally turning in the opposite direction to the front wheels.

The equipment of the vehicle was completed by an anchor and a capstan which was driven from the engine by a clutch and a belt.

The engine could be started with a starting handle. This was placed at right angles to the longitudinal axis of the vehicle. It entered the hull by an orifice which could be sealed by a bolt and a flap ahead of the right rear wheel; The movement was transmitted to the engine via the propeller drive shaft with third gear engaged in the main gearbox and the transfer gear in neutral.

Specification

Weight empty :	1,660 kg.
Carrying capacity :	360 kg.
All-up weight :	2,020 kg.
Crew :	2 men.
Seating accommodation :	5. each seat is fitted with a lifebuoy cushion.
Tires :	4 single 600 × 16 6-ply plus a spare wheel mounted on the rear deck.
Fuel tank :	57 litres gasoline 70 octane.
Electrical system :	12 volts screened and waterproofed. (2 batteries of 6 volts.)
Brakes :	drum hydraulic.
Length :	4,625 m.
Width :	1,625 m.
Height (windshield erected) :	1,725 m.
(windshield lowered) :	1,360 m.
Wheelbase :	2,150 m.
Ground clearance :	0.220 m.
Engine :	Ford type GPW. 4 cylinders in line. 2,200 c.c. 60 b.h.p. No voltage regulator on the first models. Unlike the land Jeep, the vehicle has a voltmeter.
Gearbox :	3 forward speeds, 1 reverse, with 2 speed tranfer gearbox.
Maximum gradient :	45 %.
Angle of approach :	37.5 degrees.
Angle of departure :	37 degrees.
Turning circle	
on land :	11.12 m.
on water :	10.98 m.
Maximum range	
On land :	400 km.
on water :	70 km maximum.
Maximum speed	
on land :	95 km/h.
in water :	8.5 km/h.
Freeboard	
in front :	0.430 m with breakwater.
at rear :	0.240 m.

Two MAV amphibians, Soviet copies of the GPA during a march past, showing the different shape for the rear of the hull.

Copies of the GPA

The Russian army received most of the GPAs and liked the vehicle, which proved satisfactory on the vast plains of Eastern Europe for the reasons already given. Consequently an almost perfect copy was put into production in the early fifties by the Molotov plant at Gorki. However, the axles were not partly enclosed in the hull as those on the American Jeep and the propeller drive shaft was not mounted in a tunnel but was protected by a curved steel guard. The circulation system for the cooling air was also slightly different.

The mechanical elements came from the light cross-country GAZ-69 which was made at the same factory. The new version took the name of GAZ-46-MAV or Malinki, meaning *"little amphibian vehicle"*.

In Czechoslovakia, Tatra developed a range of cross country vehicles in 1950 — 52 under the designation T-800. In this series, T 801 was a prototype for a light amphibian vehicle which, at least from the outside, bore an obvious resemblance to the GPA. The T 801 was driven by a six-cylinder petrol engine of 3,650 c.c. delivering 90 b.h.p. but it never went into production.

Specification of the GAZ-46

Engine :	4 cylinders in line gasoline 55 b.h.p.
Gearbox :	3 forward speeds, 1 reverse with 2-speed transfer box.
Length :	5.06 m.
Width :	1.73 m.
Height over windshield :	1.70 m.
Tires :	single 7.50 × 16.
Kerb weight :	1,980 kg.
Seating capacity :	5.
Maximum speed	
on land :	85 km/h
in water :	10 km/h.
Wheelbase :	2.30 m.

Civilian use of the Amphibian Jeep

If its usefulness as a military vehicle was debatable, its usefulness in the civilian sector was even more so. Apart from some enthusiasts who were looking for something unusual and did not mind the element of risk, the vehicle found hardly any users. The only services who showed any interest in it around the world were fire brigades which acquired a fair number for rescue operations in case of floods. Unfortunately, with the handicaps of insufficient carrying capacity, a body which was not very practical and insufficient power to overcome strong currents in flood waters, the GPA rarely showed up at its best.

The Tatra, Czechoslovak copy of the American amphibian.

An MAV of the Polish army with eight people on board rescuing people marooned by floods in Katowice in July 1960.

To handle the many problems posed by the presence of American troops in France, mixed Franco-American police patrols were organised like this one in Marseilles, September 1945.

THE JEEP AFTER THE WAR

Two MPs operating against the black market in Paris using one of the first types of radio telephone used by American Forces in France, with a light aerial fixed to the side of the body (October 1945).

General Leclerc and his armoured Jeep called Tailly like his tank, from the name of his family estate. The two armoured types shown here were the most common in Indo-China. The aperture in the armour plate on the second permitted the installation of a machine gun of 30 or 50 calibre in front. (Indo-China 1945).

In Indo-China in 1945 the French Army suffered a chronic lack of equipment which forced it to improvise incessantly. This Jeep carries a 30 calibre machine gun on a normal land tripod.

The British Army parachuting a Jeep during maneuvers at Netheravon in October 1947. The army ceased using Jeeps in Britain soon afterwards.

An accident between an Opel and a Jeep at Heidelberg in February 1945 provides a view of the underside of the MB with the plate protecting the transfer box.

Jeeps disembark from an LCT during operations on Bantray Island in Indo-China in 1947. They have two French M.A.C. 7.5 machine guns in front, one at the rear.

Armour plate and bullet proof glass to protect a Jeep driver.

Seen at the Lien-Chien station at the foot of the Col des Nuages in 1947 these were two of the eight armoured Jeeps converted into railway trolleys to protect communications for the French forces in Indo-China.

A Jeep converted into an armoured railway trolley on a mission to clear the lines during actions in Indo-China in 1947.

An armoured Jeep in Indo-China inspired by the war-time SAS vehicles. It was converted by the colonial ordnance establishment responsible not to the War Ministry but to the colonial administration. Operation and firing accuracy of the automatic rifle were tested at Issy-les-Moulineaux in May 1947 and although the gunner had to kneel with little chance of bracing himself against the recoil, results were satisfactory. The simple vertical gun mounting was inside a big trunk measuring 1.60m × 0.48 × 0.39.

During the first war against the Arabs the Israelis improvised armoured vehicles like this completely armoured Jeep, a rare specimen in view of the weight of the armour.

Testing a metallic carpet driven by the vehicle itself which allowed the Jeep to travel over sand or soft ground. Similar tests were carried out with a DUKW and 2 $\frac{1}{2}$ ton trucks in 1948.

United Nations observers in a white Jeep checking observations of the precarious second cease fire between Israelis and Egyptians in September 1948.

Armoured Jeep with 30 calibre machine gun of Vietnam Light Infantry in Tonkin, 1950.

Major Applegate, inventor of an electro-magnetic device to remove bits of metal from airfields to reduce damage to aircraft tyres, demonstrates it to Messieurs Duval, Duraton and Bertin of the French Thomson-Houston company which built it for the US Air Force (Orly 1949).

The Royal Air Force formed special rescue units for airmen forced down in the desert. These units seen in Egypt in 1950 comprised Jeeps, an Austin K2Y ambulance and Bedford trucks carrying food and water.

VILLEROY
I IC 14456

The South Korean Army converted some Jeeps into ceremonial vehicles with doors, extended sides, hand grips etc. This one was photographed on May 9 1954 at the reception organised for French and South Vietnam officers by the 3rd Corps of the South Korean Army.

Flying American and South Korean flags, American reinforcements arrive in the Country of The Calm Morning which can never have found its name so inappropriate.

Jeeps used in Korea for patrol and reconnaissance duties carried armoured shields. This type of protection, fairly common during World War II, was again used on the South Vietnam CJ 3 and the Military Police M 151 in Vietnam.

In Korea the same methods for carrying casualties on Jeeps were used as during the Second World War with the same discomfort for the victims (March 1951).

In December 1950 during the retreat south of the 38th parallel a Provost Marshal of the American Military Police escorts a column of Korean refugees. The Jeep has an enormous siren and flashing red light giving it priority on the highway.

A communications Jeep setting up a telephone line in a Korean valley.

Happy photographers of the French Air Force in their Ford in Indo-China in 1951.

Jeep crews often complained of the mud thrown up by their front wheels and many rigged up shields like these to protect themselves.

During public transport strikes in Paris in March 1951 Road Traffic Regulation units of the French Army organised services by military trucks between the city and suburbs. To inform passengers and control traffic, Jeeps were stationed at key points like the Opera and Place de la Concorde.

This Ford like all vehicles used for driving instruction by the French Army has a white band accross the bonnet (1951).

Lt. Frederick Dawson and his men of the 18th Infantry Regiment of the 7th US Army in Germany, invented this mounting for a 75 SR gun on a Jeep in 1952 but it was not generally adopted.

The Belgian Air Force equipped its aerodrome protection units with Jeeps inspired by those of the SAS whose exploits it had admired during the war. With the spare wheel in front of the radiator, headlamps are moved out to the wings. These vehicles like the Minerva Land Rovers which replaced them carried twin 30 machine guns behind an armoured shield in front and a machine gun of the same calibre at the rear. (Brussels 1962).

A French Air Force Ford on the high plateau of Indo-China. The front bumper which was too light for towing or for pushing broken down vehicles, trees, or other obstructions has been replaced by a steel beam. Sections of railway line were often used.

Having put into service the recoilless 75 on a Jeep, the US Army tried a 105 SR gun on an MB but it was abandoned in favour of the 106 SR.

An Air Police Jeep on duty at Orly where some of the facilities are still used by the US Air Force. The front bumper is modified and the rigid body is a familiar type although never officially adopted by the American Forces.

After the war the famous Hungarian engineer Nicolas Straussler devised these inflatable wheels to enable the Jeep to travel on water. The big wheels provided sufficient buoyancy and acted as paddle wheels.

Generals Navarre, Coigny and Gilles inspecting the site chosen for the future fortified camp at Dien Bien Phu on November 29 1953.

The US Navy tested very low pressure Rolligon tyres on several vehicles including this one based on a Jeep. The tyres were 1.20 M wide and 0.60 M high but were not very practical as they were very fragile, unsuitable for road use and considerably increased the turning circle (Lars, 1963).

In 1955 the French Army displayed at the Mailly camp the arms developed by the Technical Services since the end of the war. Among them, the AMX 13, AMX 50 and this triple mounting for Brandt 120 mm. rocket launchers on a Jeep.

Various armies experimented with radio controlled Jeeps. This one was developed by the French Army Ordnance Corps in Morocco in 1956.

North Vietnam members of the Armistice Commission travelled in Jeeps carrying on their windscreens inscriptions showing that they had been captured at Dien Bien Phu on May 7 1951.

Displays given by Italian regiments after the war often featured stunt driving on Jeeps or a blind-folded driver guided by radio through an obstacle course.

Demonstration of a French mine detector in 1954. The device could be stowed vertically attached to iron uprights on the front of the Jeep.

A feature of this model was a cutaway in the armour plate for the mounting of a 24/29 automatic rifle behind a shield in 1957.

The Jeep has always been the transport for United Nations observers in the Middle East. These are some of the Danes from the contingent sent out in 1956.

Some Jeeps were fitted out as convoy escorts in Algeria with armour protecting driver and front passenger. This one, seen in 1956 has the hinged upper panel erected.

An armoured Jeep of the same type used by the Gendarmerie leading a column of M8 armoured cars in 1956. The top section of the plating is lowered.

More blue helmets in the Sinai desert in January 1957. This time, at El Arish they are Yugoslavs as indicated by the flag painted on the body.

In Algeria the railway system was patrolled by armoured Jeeps with railway wheels made in the local workshops.

As these Jeeps could not use all gears in reverse they had to be turned round on small turntables. Like this one seen in 1957.

One of the original 25,808 MBs still in service with the French Army in Morocco in 1958...

(Below Left) This "AFN gun mounting" could be fitted to Jeep or VLR Delahaye to take a 24/29 light machine gun or a Browning B.A.R. (Algeria 1958).

A French Air Force Jeep for maintenance of the electrical system on a Thunderstreak fighter-bomber.

Although rarely used, there is a French Army conversion kit to carry stretchers on Jeeps. It is mainly used by female volunteer cadets in the Medical Corps.

The Ivory Coast Army like those of several new French speaking African States started off with equipment provided by the French Army. This Jeep is being used for driving instruction.

In 1954 when Austria reconstructed its armed forces it received equipment from various sources, but chiefly from the USA. Jeeps were supplied in large numbers in view of the need for a light cross-country vehicle in a country with such a mountainous landscape as Austria. Communications vehicles usually had closed bodies either in fabric or sheet metal and unlike combat vehicles they carried direction indicator lamps.

The Austrian Army received 106 SR guns mounted on the MB or GPW whereas these guns were at that time mounted on the M 38 or M 38Al for the US and the Allies. Equipment was exactly as on American Army Jeeps, including the two-piece windscreen and the fabric backrest for the gunner. The gun mounting was an M 79.

The agricultural version of the Delahaye Delta in 1949 before it was demonstrated to the Technical Services of the French Army.

Two Delahaye prototypes, the Delta in the background and the VLTT in the foreground, were demonstrated by the Army during the fair at the Armored School Saumur in 1952.

THE JEEP AND THE FRENC

In the immediate post-war period France was quite well provided with military motor vehicles at least as far as numbers were concerned. In addition to vehicles supplied by the Allies, the French army had captured a large number of German or Italian vehicles and in some cases Japanese.

In addition, it had been possible to restore to use a number of French vehicles which had been camouflaged during the war, some of which had been recaptured from the enemy or seized from the stocks which they had accumulated during the occupation. Finally, the national production of motor vehicles had been restarted in the face of enormous difficulties.

Unfortunately most of this material was a mixture of types worn out by several years of use and to make matters worse, lacking spare parts. For this reason there was a vast sorting operation to retain only those vehicles which still had a potential for useful service. After a review of the war surplus available a little more than 10,000 Jeeps in service or in stock were allocated to the French Army.

In 1947 the General Staff launched a programme for the design of a new generation of tactical vehicles partly to help in restarting economic activity and partly to make up the losses caused by the war in Indo-China and to re-equip the French Army. The Directorate of Armament Design and Manufacture, which was given the task of carrying out this programme, encountered great difficulties during its technical consultations, particularly with the two big national manufacturers Renault and Citroën, who declared that they were too busy restarting their production of cars for the civilian market.

In the end, only Delahaye agreed to take part in the class for light reconnaissance vehicles, whith their Delta. Presented in 1949, the Delta already had the general appearance and basic characteristic of the production models. It was a very sophisticated vehicle and technically very much in advance of its time.

The four-cylinder engine with overhead valves operated by pushrods was in light alloy with dry sump lubrication which is ideal to ensure efficien lubrication in cross country operations. It produced 63 b.h.p. at 3,800 r.p.m. A gearbox with four forward speeds, all synchronised, and reverse was in unit with the reduction gear but unlike the US Jeep, it had one lever to engage the drive to the front wheels and engage the low transfer gear. Another small lever was provided solely to lock the differential. Finally the suspension was all-independent and had a levelling system working on the rear torsion bars. So, with an o.h.v. engine, dry sump lubrication, four-speed gearbox, independent suspension and locking differential, Delahaye had adopted all the solutions thought necessary to provide their VLR with unbeatable cross country capabilities but they pursued their search for perfection into the small details. Conscious of the numerous electrical short-comings of the US Jeep,

The Peugeot 203R agricultural vehicle of 1950 met the same requirements as the VLR Delahaye. It had a 203 engine.

RMY

The VLR Delahaye on test at Satory in December 1950 shortly before its adoption as successor to the Jeep.

The Cob, final development of the Delahaye VLRD which was only tested briefly before the manufacturer closed down.

The Peugeot 203 Type RA had a much more martial appearance than the R Type. Its engine was fitted with a 403 cylinder head and was rated at eight fiscal horsepower instead of the seven for the 203R.

they broke new ground by using 24 volts. The coil and distributor were mounted as high as passible on the engine to give maximum protection against water splashes; the fuse box was accessibly mounted on the instrument panel. Everything was thought out and designed with a view to use over all kinds of terrain.

After several trials, the bodywork was modified; windscreen in two sections, bonnet opening towards the front, new design for the side panels of the body and for the radiator grille which now incorporated the headlamps. The vehicle was named VLR-D for Voiture Légère de Reconnaissance Delahaye. The Army kept on testing it mercilessly but with expert drivers, at Satory, Monthléry, Bourges, on the pavé in the north, in the Alps and in the rally from the Mediterranean to Cape Town, all of which the VLR survived.

After further modifications to the body, the VLR 4 × 4 Delahaye quarter ton model 51, entered into service with bodies made by Facel and to its cost it then fell into the hands of inexpert drivers. Problems soon arose. The Army got its VLRs, certainly, but they received no technical publications on driving and maintenance and this gave rise to various mechanical problems. The VLR went well, too well. It was more

Peugeot VSP, the company's reply to Delahaye's Cob prototype. Ten pre-production examples ware built and tested but no orders resulted.

Another possible Jeep replacement was presented by Fiat/MAN/ Saviem. Although it was not adopted Germany gave it extended trials in service. It is seen here during amphibian tests in Corsica.

comfortable than the US Jeep and rather faster, so its drivers became over-confident and serious accidents resulted.

The road holding was blamed and it was alleged that the differential locked itself. The truth was often different. Though the differentials did lock up by themselves on rare occasions, the drivers often locked them inadvertently (on the US Jeep the second lever engaged the low transfer gear.) or simply forgot to unlock them and check the fact after covering a difficult section.

The accidents involving the road holding could also be explained if one considers these three points; a geared up fourth gear, independent suspension with a long travel and variable setting for the rear torsion bars.

But there is no smoke without fire; though the VLR was technically advanced it was not ready to be put into the hands of just any driver. Many modifications were made to try and improve the VLR, which developed into the model 53, but in vain.

Delahaye recognised these problems and started to develop a new prototype, the Cob. There was just time for the Army to give it a preliminary test in 1954 before Delahaye was absorbed by Hotchkiss, who chose to play the Willys card. In any case, the

When France, Germany and Italy agreed on a joint programme to find a Jeep replacement, Hotchkiss/ Lancia/Bussing proposed this model with a BMW engine in 1970. Heavy and complicated, it was not ordered. One of the prototypes is in the Armoured Vehicle Museum at Saumur.

Enterprising spirits devised some surprising solutions like these prototypes built by the French Navy on the basis of the Citroen 2CV van with 75 SR or 20 calibre guns.

In 1964 Victor Bouffort designed this prototype called the Bison VB 100 and had it built by Batignolles-Chatillon Locomotive Works.

military users had let it be known that they would turn their attention to simpler vehicles in future.

In all 9,632 VLR -Ds were built for the Army and certain civilian users, the latter for the most part being equipped with 12 volt electrical systems.

Peugeot were perhaps less convinced of the interest of this programme and therefore did not take part in it directly but carried out their own studies independently, so it was not until August 1950 that they presented their Voiture Agricole Type 203 R for homologation by the Service des Mines. At the same time tests were conducted by the military authorities at the Allondans military ground.

Although very similar to the US Jeep in appearance and general conception, the 203 R was none the less a real Peugeot. The engine was the 1,290 c.c. (75 × 73 mm) unit of the 203 with pushrod-operated overhead valves, the gearbox had four forward speeds with a geared-up top, the transfer box was identical to the American one, with two speeds and provision for engaging front-wheel drive. Front and rear axles were typically Peugeot with their worm final drive (worm above the crown wheel).

The light weight and reasonable cost of the Citroen Mehari led to its widespread use by the French forces for liaison work.

The Marine Commandos used some of these Renault R4 Sinpar 4 × 4 with special fittings for dropping by parachute.

The Commandos at Lorient used the Renault Sinpar 4 × 4 and several were embarked on the helicopter carrier Jeanne d'Arc.

In February 1973 Panhard designed a light vehicle called the M7 with 250 kg carrying capacity, bodywork by Heuliez and 1222 c.c. Citroen GS engine. Two versions were planned; military 4× 4, civilian 4× 2.

Although from the military point of view the 203 R was handicapped by arriving later than the Delahaye, Peugeot nevertheless continued to improve the vehicle. They put the fiscal horsepower up from 7 to 8 by installing an engine of 1,460 c.c. (80 × 73 mm) and made it mechanically more suitable for agriculture by adding a power take off, and in May 1951 they presented it as their Voiture Agricole Type 203 RA but this one had no better luck.

It was only after the many problems encountered by Delahaye became known that the Peugeot project again emerged to take its definitive form in December 1954 with the Type 203 RB agricultural vehicle and as the Army was once again looking for a light all-terrain vehicle, Peugeot tried their luck again with a military version of the 203 RB, the VSP (Vehicle Special Peugeot). After satisfactory tests of the prototype a pre-series of ten VSPs was ordered to allow the STA Technical Section of the French Army and the AMX Ateliers d'Issy les Moulineaux-a military testing establishment, to make a more thorough investigation. Intensive testing began in 1956 and revealed numerous defects.

The Peugeot engineers remained confident and put them right. The rear axle was improved. The liveliness of the vehicle was considerably improved by fitting a cylinder head from the 403. Electrical screening and suspension were reviewed. During the whole of 1957 and 1958 the military engineers and those from Peugeot worked incessantly to improve the VSP. Technically it had become a sound vehicle, appreciably better than the VLR Delahaye, but for all that its military career ended there, and the vehicles were returned to Sochaux where they were handed over to the factory fire service.

So the Peugeot solution was unfortunately never put to the test but it did demonstrate that it is technically feasible, at least in this category, to create a specialised vehicle using a high proportion of mechanical elements which are already in mass production. *"But if it had been adopted, an important manufacturing problem would have arisen; it would have been necessary to set up a highly efficient commercial production line for a limited period of one or two years in order to gain the advantages of mass production, because the components concerned were derived from civilian production which was in a state of constant evolution."* (Lt. Col. Perrodon, Revue de l'Armée No. 21 Aug. 1962).

These manufacturing problems confronted the General Staff once again when they sought a replacement for the M 201. Few cross country vehicles were then being developed in France as the requirements of the principal user were met by the M 201. Without a substantial order which only the armed forces could provide, any attempt at manufacture even in small numbers was doomed to failure. All the same one should mention the light communications vehicle with two-wheel drive which was developed in Morocco by Georges Irat. Called the Voiture du Bled it was the subject of a number of tests by the military technical services.

While awaiting the result of the latest competition for a Jeep replacement French manufacturers continued to offer ideas like this Citroen A 4×4 seen at the Demeter maneuvers, 1979.

Competitor for the Mehari on a Renault R4 or R6 platform, the Rodeo built by ACL can be fitted with the Sinpar 4×4 conversion. Some of them are used by the French Navy. (1978)

TO REPLACE THE JEEP...

When Mr. Henry Ainsworth, Managing Director of Hotchkiss and Director of the company's automobile division returned to France after the second World War, be brought with him one idea; the Jeep.

Mr. Ainsworth, who was a British citizen, had escaped to London at the beginning of the war. Because of his capabilities in the automotive field and particularly his experience in the specialised task of building the H 35 and H 39 tanks which Hotchkiss, under his management, had supplied to the French Army, the British war cabinet attached him to the British purchasing mission in the USA. His particular job was to find manufacturers for the armoured vehicles ordered by Britain from the American industry.

During the same period, the French purchasing

Two views of the Hotchkiss Jeep M 201 ready for delivery to the French Army. Except for minor details it is the American Jeep but made in France. There are no wooden rests on the bonnet for the windscreen but there are guards over the wind-screen wipers, and there are two radio aerial mountings.

HE JEEP!

mission directed by M. Jean Monnet had, with the support of the last government of the Third Republic, signed a protocol of agreement with the British which allowed the British to receive the supplies ordered in the USA by France under the Cash and Carry agreement.

While liquidating the French contract, Mr. Ainsworth met Mr. Marcel Müller who was on the staff of the commercial attaché at the French Embassy in Washington and the two men became friends as the war continued. M. Müller eventually entered active service as an officer in the American Army and he opted for American nationality soon afterwards. After being demobilised, M. Müller entered the American motor industry where after some false starts, his career took a definite turning with Willys-Overland.

The Hotchkiss assembly lines for the M 201. Chassis and bodies were fitted out separately.

Their friendship led Henry Ainsworth to propose to the Board of Directors of Hotchkiss in 1946 that he should take over the sale and distribution of the Jeep Universal Type CJ 2A and then CJ 3A in France and the French colonies as this looked like a viable operation which could enable the Company to start up in business again. The Directors accepted the offer and responsibility for commercial operation involving the vehicle was entrusted to SOFIA (Société Financière et Industrielle Automobile).

This company, in which Hotchkiss had a majority holding, had been formed before the war to administer the assets of Amilcar when this company was taken over by Hotchkiss. During the hostilities, the buildings had been used as a garage by the German Army and then at the liberation were taken over by the American Army for the same purpose. As official distributor for Willys-Overland, SOFIA organised demonstrations to show the advantages of the Jeep.

The small team added to its numbers day by day and recruited a Dutchman, a veteran of the Princess Irene Brigade, M. Sanders, who was to become one of the principal architects of the company's success and commercial director of Hotchkiss. There was also a native of Biarritz, Jean-Marcel Mauroux whose wonderful Basque accent soon became familiar to al Jeep clients in France.

However, as yet this activity only concerned th Jeep Universal CJ 2A and CJ 3A and B of the Willy civilian range and no French content was the planned for the vehicles, which were imported com plete from the USA.

The first real Hotchkiss contact with the MB mode took place in 1953 when the company had to execut several orders for spare parts; engines, axles, etc. either for the French Army or for overseas, in conformity with an agreement signed on June 15 1952 when Willys granted Hotchkiss the right to organise the market for spare parts for war-time Jeeps. This was the first time that quantity production of spare parts for the MB had been undertaken in France since the Simca plant, then under the management of J.-A. Grégoire had rebuilt 75,000 Jeep engines for the American Army in 1944/45.

Also in 1952, Willys granted Hotchkiss a manufacturing licence for France and the French colonies, covering the Jeep Universal, its spare parts and components as well as all other types of vehicle derived from the Jeep family built by Willys Overland. Although they sold hardly any, Hotchkiss quoted in their 1962 price list the prices for the station wagon and the forward control light truck FC 150 or 170, etc. proving that St. Denis had a genuine intention to sell not only the Jeep but the whole Willys range even though they had no success in France. Hotchkiss was then the only company in Europe to have such a licence.

This agreement also allowed Hotchkiss to sell to countries which were short of dollars like those in Africa and which consequently could not obtain supplies from the United States.

The first civilian Jeep to be assembled in France, the JH 101 was quite close in its specification to the American CJ 3B and still incorporated a high proportion of imported parts; engine, gearbox, transfer box and axles. 3,496 units were built up to 1960 when

Body drop on the Hotchkiss M 201 and fitting of the wheels. Chassis frames are seen in the background.

Engine compartment on the 24 volt M 201. All electrical equipment is screened. There are two 12 volt batteries and the voltage regulator takes the place of the air filter which is moved close to the left wing.

Jeeps of the 21st RIMA at Sissones. In the foreground the sheet metal guard over the windscreen wipers is visible and further back are Jeeps fitted with the 106 SR gun. (June 1964).

A disembarkation exercise. On the bumper is a case containing the camouflage netting and on the wing the supplementary air filter included in the Sahara kit for protection against sand.

A captain in a delicate situation with his Jeep. Not an occasion for spectacular driving.

231-0416
031 732
231-0415

the vehicle was replaced by a model of entirely French manufacture, the JH 102. A long chassis version of the JH 102, the HWL, emerged in 1963.

Hotchkiss was also the first Willys licensee to offer a production model of the Jeep with a diesel engine, in this case the Indenor 85 XD P4 in 1962, although many private owners had converted Hotchkiss or MB models, mostly using Perkins engines.

The commercial arrangements were modified when the parent company took over control of this sector, SOFIA remaining a priveleged distributor controlling the distribution of spare parts in France and the sale of Jeeps in the Paris region.

The Jeep always attracts children. This one has extensive radio equipment.

Jeeps can be dropped by parachute on a suitable mounting after special preparation. This one in Algeria in 1959 rests on straw bales which serve as packing and soften the landing.

Easily carried by air, the Jeep can intervene quickly. This one is emerging from a Super Frelon helicopter.

Reviewing the French troops in the Sahara in 1961. The direction indicator lamps in the grille are unusual on a military Jeep. In front is the box for the camouflage net. On one wing is the Sahara air filter and on the other a siren.

The year 1954 was notable chiefly for an important rescue operation when Hotchkiss took over the assets and liabilities of the Société des Automobiles Delahaye, an operation approved by extraordinary general meetings on June 9 and July 29 1954. Hotchkiss carried the sales of VLR vehicles and spare parts through to their conclusion. It was moreover, under the auspices of Hotchkiss that tests were carried out on the Cob, latest and last offshoot of the VLR line.

Under the new name of Société Hotchkiss-Delahaye the amount of effort devoted to the Willys Jeep was considerably increased. In 1955, Hotchkiss installed manufacturing facilities and a production line in the Ornano plant at St. Denis which made possible series production of the Hotchkiss-Willys Jeep Type JH 102 from 1956 onwards.

In 1955, after what must be regarded as the failure of the French programme for a light cross-country liaison vehicle the Army decided to re-start manufacture of the MB. This decision has often been viewed with astonishment. Why order a vehicle 15 years after its conception when new models had already replaced it on the manufacturer's own production lines? Two replies have been given to this question.

In the first place, the models built by Willys in the fifties differed only slightly from the war-time version

For maneuvers in open country, a natural coating of mud is often a sufficient camouflage.

Drive past by female volunteers of the medical corps of the French Army Cadets on stretcher-carrying Jeeps.

and it was easy to incorporate a proportion of the recent improvements made by Willys into the MBs built by Hotchkiss.

Moreover, the French Army had large stocks of spare parts and nearly 10,000 vehicles in service either via Lend-Lease or purchased as war surplus from the USA. It was thus an opportunity to do a good deal from the point of view of fleet management while acquiring a well proven vehicle that could easily be rejuvenated if desired.

Finally, it should not be forgotten that the attempt to put into service a 'modern' vehicle had already been tried with the VLR - D and after this failure a simple and well tried vehicle could not fail to appeal to the military. This opinion was further reinforced by the failure of the Austin Champ in England where as with the VLR, it had proved impossible to build and keep operational a light all-terrain vehicle incorporating the latest developments in automobile design. So it seemed that there was nothing to do but buy the MB once more because all attempts to replace it had failed as soon as anyone departed from the technical solutions incorporated in the original design.

In this situation, it seemed better to put up with a

A Jeep on reconnaissance during maneuvers with the French Army equipped with an AA 2 machine gun.

A Jeep with infra-red driving and aiming equipment for night operations. Note the special driving lamps and the goggles worn by the driver.

An M 201 with a 20 M 621 cannon controlled by a gunner at the side of the vehicle. This prototype was presented at the Satory exhibition but never went into service.

M 201 with launcher for wire-guided SS 10 missiles. Few of them entered service, as the ENTAC did substantially the same job. Some of the MBs and GPWs carried also the SS 10. But the regular mounting was on the M 201. Some mounting like this one were modified to test fire the SS 11.

M 201 fitted with ENTAC missiles ready for operation during maneuvers in Africa.

An ENTAC Jeep with special hood frame. Mounted in front is the wheeled transporter for carrying the missiles on the ground which was very rarely used.

On December 31 1969 all production was stopped at what had then become the automobile division of Thomson-Houston Hotchkiss-Brandt, a company which had been formed in 1961 with M. Richard as President. The production line making the JH 102 and HWL at the Stains plant was moved to the VIASA works in Spain which after various modifications went on building these models for some time.

Technical Specification

The M 201 is generally similar to the war-time Jeep in its bodywork and mechanical components but there are a certain number of items differing between one model and the other.

The Engine

It is commonly said that the engines are identical except for the type of distributor. This is partly true, partly untrue. It is true that almost all the engines of the war-time Jeeps had a chain-driven camshaft. It is untrue in that contrary to what is generally believed,

fairly simple vehicle which, all things considered, did its job, rather than go running after what looked like a technological will'o the wisp. An initial order was therefore issued for 465 *«Jeep Hotchkiss licence MB»* vehicles.

By the time production ended at the Stains plant in 1966, 27,628 M 201 had been built of which 27,604 were sold to the French Army and 24 were acquired by other departments. The Direction de la Securité Civile i.e the Civil Defense Corps (then called the Service National de la Protection Civile), a branch of the Ministry of the Interior, bought 15 of them.

Besides these 27,628 M 201, there were 5,554 civilian Jeeps built between 1954 and July 1969 of the JH 101, JH 102 and HWL models and some hundreds of CJ 5 and CJ 6 and some special models imported directly from the USA.

The Oto Melara 105 mm. Howitzer used some time ago by artillery units of the Parachute Division and by the French Alpine Brigade is light enough to be towed by a Jeep.

This Jeep carries eight ENTAC wire-guided missiles, double the normal complement. Although possible with standard mountings the arrangement was not adopted by the French Army.

A well camouflaged Jeep with a Milan missile launcher during test firing.

Milan Jeep on show. There are three containers for missiles at the rear.

Colonial units of the French Army were the last to use 75 SR guns on Jeeps as this weapon was no longer powerful enough by European standards.

the engines of the last series in 1945 were fitted with the new gear drive to the camshaft which was introduced on a large scale in 1946 on the engine of the CJ 2-Universal.

Naturally, the drive on the M 201 engine is by gears. On the M 201 Hotchkiss-Brandt engine the cylinder head is reinforced.

Various items of equipment are, of course, French. The carburetter is a Solex, the model varying with the year of manufacture. The cooling system is identical except for the water pump; (twin-throat pump on the M 201 with 24-volt electrics). The radiator on the war-time Jeep has a cowl which is absent from the Hotchkiss.

Universal Joints

Hotchkiss never used the Rzeppa joint. The fitting of the Bendix joint on the M 201 differs from that adopted on the American model in the absence of screw threads.

Transfer Gearbox

Here also the M 201 borrows from the CJ 2 its larger shafts and they are mounted on roller bearings whereas the Hotchkiss under MB licence has the same shafts as the wartime vehicles mounted on needle roller bearings.

Bodywork

Although the method of supporting the folded windshield by two metal handles fitted above the glass is theoretically a Hotchkiss identity feature, it is not entirely to be trusted because the first series retained the two wood blocks on on the bonnet originally provided for the same purpose.

The M 201 fitted with the 106 mm recoilless gun had a different windscreen, allowing the gun to pass through with the screen erected. The rear panel of the body was cut away, the rear cross member of the chassis extended and the spare wheel mounted on the right side of the vehicle, which made it necessary to bend the exhaust system downwards to avoid damaging the tyre. On the other hand the French-built models all had electric windshield wipers mounted above the central upright on the windshield.

Suspension

The springs always had ten leaves at each side in front on the French model and 11 leaves or more at the rear, according to the use of the vehicle. The 106 SR gun, fire service, Entac and Sahara had

106 recoilless gun on a Jeep showing the substantial tripod mounting.

The long barrel of the 106 SR required a slot in the windscreen offset to avoid encroaching on the space for the driver.

Jeep with 106 SR in action. It also has infra-red lighting.

There were two types of windscreen obtained from American conversion kits for Jeeps fitted with the 106SR gun.

Jeep equipped with prototype mine laying equipment designed for use by the French forces.

reinforced suspension and stronger shock absorbers, usually by Katz.

The American Jeeps had nine leaves each side at the rear and eight leaves set up differently in front. In addition, a compensator spring was added in course of production (see the chapter on the American Jeep).

Handbrake

One of the defects on the Jeep was the inefficiency of the handbrake. Only its simplicity had justified its adoption in production. Therefore on the last models of transfer gearbox built during the war the handbrake was improved by replacing the brake band with a drum brake. This was also carried out as a retrospective modification in the workshops of the American Army after the war. Although the responsibility for this modification is generally credited to Ford, it was fitted as standard by both Willys and Ford. Hotchkiss adopted this improvement for their handbrake.

Electrical System

The two types of Hotchkiss military Jeep had electrical equipment of 6, 12 or 24 volts screened or not according to the date and the market they were intended for. However, of the vehicles now in service, most have 12 or 24 volt equipment with screened wiring.

M 201 with prototype protective bodywork designed for use by the French Gendarmerie. There is metal netting over windows and radiator.

Sahara Equipment

Special equipment on the M 201 Sahara consisted of a supplementary fuel tank under the front passenger seat, an additional electric fuel pump under the front floor on the left of the body and a Tecalemit first stage air filter on the right wing.

The operating equipment included a mat for extracting the vehicle from sand. A galvanised Jerrican was fitted between the front seats, clips and eyelets being added to the bodywork.

Tyres

The tyres were 600 × 16 except for the Sahara model which could be fitted with 6.50 × 16 or 106 SR in 6.50 × 16.

For the Saumur carrousel some M 201 vehicles were equipped as mobile ramps on which motor cycle recruits could demonstrate their sense of balance and their sang froid in riding exhibitions.

Line where Jeeps were dismantled. Everything was completely stripped and rigorously inspected.

Sub-assembly workshop. The chassis in the foreground is a test rig for reconditioned axles.

LA MALTOURNÉE

Among the Ordnance Service units and establishments which provide the equipment back-up for the French Army are some assigned to special tasks. This was the case with the Etablissement de Reserve Generale du Materiel Automobile, the general motor vehicle reserve depot at La Maltournée in the Paris suburbs which rebuilt light cross-country vehicles, meaning mainly Jeeps, between 1946 and 1979. The site and the buildings, which dated from 1900, had been bought from Thomson-Houston by the Ministry of War in 1938 to form an annexe to the central automobile depot at Vincennes.

The annexe was occupied by the Germans from 1940 to 1944 and was serverely damaged during the liberation. In 1945 reconstruction was started and a workshop for vehicle repairs was installed. From that date the workshops, which were administratively attached to the regional ordnance depot at Vincennes (up to 1949) became the ERGM/AU of La Maltournée.

It developed into a big business with modern equipment and 350 workers. It had several tasks, the main one being to rebuilt light cross-country vehicles which were worth repairing but were beyond the capacity of the local workshops.

It also undertook the rebuilting of assembles and sub assemblies on behalf of the central supply services.

It reconditioned components or complete vehicles, units and sub assemblies for long term stocks.

It carried out tests and experiments and made returns on behalf of the technical inspectorate and particularly for the central supply services.

It maintained vehicles and machinery issued to army units.

It kept stocks and maintained equipment for mobilisation (equipment used by the engineers and by the Army Service Corps for the medical and fuel supply services.)

Each year the Central Ordnance Division fixed the

On the Jeep assembly line the reconditioned chassis receive their rebuilt engines.

welders repaired them. Repained and checked dimensionally these parts were then ready to join the assembly line.

Units and mechanical sub assemblies, engines, gearboxes and axles were cleaned and then sent to the various workshops to be dismantled and checked by specialised personnel.

The engines were dismantled by a real virtuoso; in fact practically all the Jeep engines rebuilt at La Maltournée were dismantled by one and the same workman. After sorting, the parts judged re-usable were repaired and put into the supply circuit for the engine assembly line. Then all the component parts for an engine were put on a tray and delivered to the engine building positions. When it was rebuilt, the engine was run in on the bench and then brake tested to plot its fuel consumption and power curves. After the rebuild, each unit was given a meticulous inspection with the aid of precision equipment sometimes specially designed by the metrology laboratory

number of Jeeps to be rebuilt by the establishment in the following year. For example, the programme for 1972 included

- 552 VLTT Hotchkiss.
- 367 VLTT Hotchkiss 6 and 12 volt to be converted to 24 volt.
- 210 VLTT US Willys 6 and 12 volt to be converted to 24 volt.
- 10 VLTT US Ford 6 and 12 volt to be converted to 24 volt.
- 10 VLTT to be converted into ENTAC anti-tank rocket launchers.

Total, 1,149 VLTT.

The Jeeps to be rebuilt were delivered by rail and put into stock to wait their turn according to the plan. Then they were put onto the dismantling line. The chassis, the body and the various body fittings were sent to the sheet metal workshop and the paint shop where after then had passed through acid baths to remove the old paint the sheet metal workers and

Before and after. Results of a systematic rebuild.

ERGM/AU AT LA MALTOURNÉE

ACTIVITIES OF THE ESTABLISHMENT FROM 1946 TO 1948														
Materiel	1946 1950	1951 1955	1956 1960	1961 1965	1966 1970	1971	1972	1973	1974	1975	1976	1977	1978	Totals
VLTT Jeeps	4 941	3 208	7 704	9 998	7 020	1 238	1 149	1 161	942	795	550	363	187	39 256
VLR Delahayes		113	461											574
Jeep engines	5 726	5 076	8 273	10 652	8 424	2 401	2 592	1 886	1 485	1 447	1 305	1 197	611	51 075
Front Axles	5 353	4 353	7 607	10 042	8 183	1 217	1 128	1 343	1 373	804	541	332	166	42 442
Rear Axles	5 352	4 368	7 278	9 984	8 128	1 277	1 107	1 354	1 390	999	952	400	160	42 749
Gearboxes	5 357	6 137	8 050	10 224	13 393	1 258	1 129	3 333	3 673	3 526	1 867	1 390	172	59 509
Transfer Boxes	5 357	5 437	8 248	10 223	6 996	1 129	3 333	3 673	3 526	1 857	1 390	172	52 609	
Steering Gear	5 035	3 443	9 821	14 730	8 509	2 902	2 274	3 593	3 277	3 875	2 484	460		60 403
Delahaye Engines		940	3 387											4 327
Front Axles			83											83
Rear Axles			301											301
Gearboxes		154	359											513
Steering Gear			84											84
Citroen 6 cyl Engines		1 422	297											1 719
Citroen 4 cyl Engines	912	5 336	80											6 328
Miscellaneous Engines			23		84		45							152
Sub-assemblies SCA			32 053	95 244	3 417	13 530	17 665	5 954	18 824	10 000	6 000	7 500		205 187

of the establishment to make sure that no problems would arise in service. Having passed scrutiny, the units joined the final assembly line.

Reassembled stage by stage, just as on the original assembly line, the complete vehicle was then tested to check on its behaviour and spot any mechanical imperfections. Although each unit had been checked before assembly, road tests could still show up problems in an axle, a gearbox or an engine.

After repair or replacement of any defective item, the vehicle was tested again and if it proved satisfactory it was ready to be presented to the acceptance commission. Maltournée enthusiastically exploited the standardisation which had been adopted at the creation of the Jeep, cheerfully mixing parts by Ford, Willys or Hotchkiss in the same racks to be used on any vehicle provided they were of the same type and intended for the same purpose. One could even see vehicles with a Willys front, a Ford rear and other components from Hotchkiss. So what came out of the ERM was not Willys, Ford or Hotchkiss Jeeps but quarter-ton VLTTs. (Voiture Légère Tous Terrains) The only Jeeps to retain their identity if not their integrity were those on which only electrical work was carried out.

Eventually the rebuilding of assemblies and sub assemblies on behalf of the central supply services saturated the workshops. For example the Ministerial programme already quoted for 1972, in addition to the reconstruction of complete vehicles, provided for

Units

- 95 engines VLTT 12 volt
- 500 engines VLTT to receive 24 volt equipment
- 750 engines VLTT 24 volt
- 198 engines VLTT 6 and 12 volt to be changed to 24 volt

Sub assemblies

- 1,151 VLTT steering gear
- 10,912 front brake shoes
- 704 rear brake shoes
- 1,201 suspension springs
- 237 distributors

This closed bodywork was specially designed for the French Air Force and built in small batches by the ERGM/AU at La Maltournée.

- 971 dynamos
- 1,122 voltage regulators
- 2,318 plug leads
- 200 carburetors

Alongside these major operations, La Maltournée also had to design and build a number of prototypes and special batches. At the request of the Air Force they produced a small batch of steel bodied Jeeps, called "Break d'Aviation" (aviation station wagon), which gave efficient protection against the elements and a degree of visibility to the side and the rear which none of the improvisations of the last war had been able to provide.

To improve the performance of the Jeeps which were specially prepared for the rallies in which the army took part, La Maltournée designed two lightened prototype Jeeps making maximum use of aluminium and devising a cylinder head in light alloy which reduced the weight and increased the compression ratio. But one of their oddest achievements was the construction of four Jeeps called Malt 1, 2, 3 and 4 which were specially designed for the *"Nuits de l'Armée"*. During these great popular fetes which were army spectacles like a British tattoo, many demonstrations were staged. Following the example of the British, one of them involved the highspeed assembly of a Jeep. As performed by the British, this consisted in building a Jeep in less than five minutes from a kit lying on the ground; in France it took the following form. The Jeep entered the exhibition ground with four men on board and could only leave after passing through a small aperture. As it could not go through complete, the Jeep had to be dismantled into units that could be carried through the obstruction and was then reassembled and driven out of the ground, all in less than ten minutes.

Although they looked identical to the normal Jeep, the Malt vehicles were very much modified. Many types of rapid fixing were developed to attach the wheels, the axles or the bodywork which was divided into two parts. The power unit was a runner but the fuel tank had been replaced by the air filter which was specially adapted, the objective being not to run for a long time but to put on a show. These four Jeeps are kept at the ERM at Versailles-Satory and still give regular demonstrations.

Other design studies were carried out for the improvement of the Jeep itself or to solve various technical problems, such as devising the reinforced suspension for the ENTAC Jeep. Various experiments were also made on the VLR Delahaye which was rebuilt by La Maltournée between 1951 and 1960. The defects of the VLR were reviewed and after lengthy efforts, solutions were discovered. The most important involved the control system for the locking of the differential, the pressure lubrication of the engine and its cooling system. It is interesting to note that although the engineers and mechanics at La Maltournée were specialistes in the Jeep, once they had cured these problems on the VLR Delahaye it no longer had any major defects and they considered it very much superior to the M 201 Hotchkiss.

The backlog of Jeeps and units for rebuilding dwindled considerably from 1974 onwards and the establishment was progressively run down, to be closed finally in November 1978. The last 80 Jeeps to be officially rebuilt were M 201 Hotchkiss with ENTAC missile launchers.

The maintenance of the Jeeps remaining in service is now carried out by regional workshops while awaiting their replacement by the new VLTT which the French Army has now ordered.

Although it looks normal, the Malt Jeep is full of surprises. The special attachments for the radiator grille are visible and the joint where the body is split into two halves.

Principal Modifications made to MB, GPW and M 201 Jeeps by the French Army

Date	Modification
January 23 1945	Reinforcement of cranked steering arm.
June 1952	Reinforcement of rear spring mounting.
June 1956	Fitting spare wheel mounting on US 1/4-ton trailer.
April 1958	Fitting Ford auxiliary gearbox to Willys chassis.
October 1960	Modification to fitting of screened plugs on M 201 24 volt.
March 1961	Fitting AN GRC 9 radio.
April 1962	Modification of dynamo mounting on M 201 to receive Paris Rhone FPRG 15 R 45.
June 1963	M 201 fitted with SS 10. Reinforcement for side of body.
June 1963	M 201 24 volt : Fitting guard for master switch.
February 1964	Modifying front brake back plates.
February 1964	Fitting carrier for synthetic camouflage net 1962 No. 1 behind spare wheel.
April 1964	Modification to drain plug on auxiliary gearbox.
August 1964	M 201 24 volt : Fitting of radio sets AN GRC 9, AN VRC 10, AN VRC 18, AN PRC 10, AN PRC 8, RR TP 2A.
August 1964	MB, GPW, M 201 6 and 12 volt : Fitting junction box for radio.
September 1964	Reinforcement of cranked steering arm.
October 1966	Fitfiring equipment for ENTAC rocket launcher on M 201.
October 1966	Fitting SEV-Marchal type SS screenwiper motor on Hotchkiss M 201 24 volt.
December 1966	Fitting equipment to carry ENTAC rocker launcher on roads and across country on M 201.
June 1967	Modification to convert Jeep for ENTAC.
October 1967	Jeeps, all types : Adapting electrical equipment to make it legal for road use.
January 1968	M 201 24 volt : Reinforcement of suspension and modifications for carriage of 106 SR.
November 1968	Mounting to carry containers for chemical decontamination.
March 1969	M 201 ENTAC : Fitting carrier for camouflage net in front of radiator grille.
November 1969	M 201 24 volt : Fitting of radio sets AN PRC 9, AN PRC 10, AN VRC 9, AN VRC 10, AN VRC 17, AN VRC 18, AN GRC 5, AN GRC 7, AN GRC 9, SCR 506, SCR 508, SCR 528, SCR GD 608, TP VP 13, TP VP 213, TR VP 14, RRTP 2.
July 1970	MB, GPW, M 201 6, 12, 24 volt : Fitting nylon top with side curtains and doors.
March 1971	Waterproofing for immersion of the M 201 radio VLTT with LTCV 4 or LTCV 5.
September 1971	Fitting of new fuel pipe on M 201 24 volt.
March 1973	Waterproofing of M 201 ENTAC.
December 1974	Modifying trailer lighting system to conform with road traffic regulations.
April 1976	Convert M 201 24 volt to carry Milan missile.

Initially the French Army tested a number of light cross country vehicles which might replace the Jeep. One was the VELTT, a French vehicle built by STEMAT in small numbers and seen here before its test. The body is in sheet iron and is loaded with pig iron corresponding to the claimed carrying capacity.

THE FUTURE

The M 201 was chosen by the French forces because no modern vehicle had proved capable of replacing the Jeep, either for technical reasons or financial reasons. Despite various attemps in the early sixties, such as the Bison by Victor Bouffort (1964), a design by Marmon-Bocquet and the order from the French Navy for the Renault R4 Sinpar with special bodywork in sheet steel (four examples in 1965, six in 1966) it did not seem possible, chiefly for economic reasons to create a French Jeep.

In fact, the Army needed some 10,000 vehicles, a number which excluded small scale manufacture by

The Land Rover Military Prototype on the 109 in. long chassis with V8 engine and Range Rover transmission. Three were built and a civilian version emerged nearly a year later.

The M 151 A2 which was tested and although not adopted, was frequently used as a standard of comparison.

Citroen was the last major French manufacturer to offer a Jeep replacement. It was the C 44, a French version of the VW Iltis.

hand yet was too small to justify manufacture on the production lines of the big car manufacturers.

The Italians and the Germans also needed a new light liaison vehicle to replace their Campagnolas and Mungas which were beginning to show their age. If they were all combined into a single programme, the requirements of the three nations amounted to some 50,000 vehicles, enough to justify launching a proper industrial programme.

Thus a tripartite programme was launched, embracing French, German and Italian manufactureres, who were invited to get together to present their proposals. The vehicle described in the design requirements naturally took into account the desiderata of each army and specified that the vehicle must be amphibian without special preparation, must be able to carry four men and their equipment — say 500 kg carrying capacity — must have sealed and screened electrical equipment and be suitable for aerial transport and for dropping by parachute. Furthermore, at

 Saviem or rather Renault Vehicules Industriels presented the TRM 500, a French version of the Fiat Nuova Campagnola.

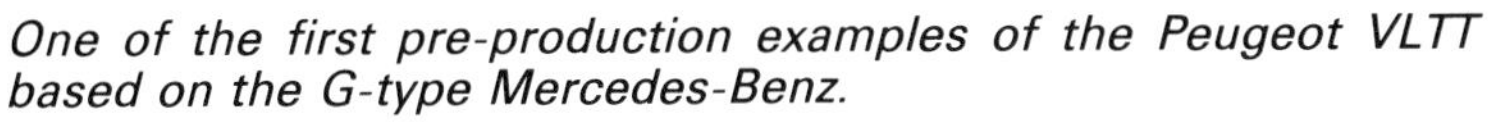
One of the first pre-production examples of the Peugeot VLTT based on the G-type Mercedes-Benz.

Rear view of the Peugeot VLTT showing tailgate and pivoted spare wheel mounting.

the request of the Germans, the front and rear axles must be identical to simplify the problem of spare parts.

The engines must run on gasoline.

Bussing, Hotchkiss and Lancia formed a consortium to compete with Fiat MAN Saviem. The vehicles were ready to begin their tests at the beginning of 1969 but the tripartite programme never got very far because of economic and financial problems. By the end of the same year, the automobile division of Hotchkiss had closed down, Bussing was taken over by MAN and Lancia by Fiat, so one of the competitors was out of the running.

In addition the design requirements were so restrictive that the price of the vehicles had increased to such an extent that the FMS was abandoned in spite of a satisfactory performance, the prototypes already built being retained by the manufacturers.

The thing was obviously becoming complicated and in the meantime, France had decided to make the

C44 on the Citroen stand at the Satory VII Exhibition in 1979.

TRM 500 exhibited by RVI in pre-production form.

Peugeot P4 during the Korigan maneuvers in Brittany in 1979.

existing stock of vehicles last as long as possible.

To do this, some Citroen Meharis were brought to replace the US Jeeps and the M 201 in work-a-day tasks. The Jeeps thus liberated were handed over as reserves or replacements for worn out vehicles in combat units.

In 1968 the first ten Meharis were delivered to the services. Ten years later, 8,807 of them had been delivered to the three armed services and to the Gendarmerie.

The failure of the tripartite programme led the French Army to start technical consultations with the automobile industry. The manufacturers (Renault, Panhard on behalf of Citroen, etc) put forward feasibility studies or designs in 1972 and 1973 but matters rested there. The requirements of the French Army had not changed, nor the industrial problem mentioned above which flowed from them. A new approach was therefore adopted, directed at the commercial sector in the hope of finding a vehicule already in quantity production which would meet the French requirements while avoiding some of the expense and delays involve in a special design.

The Etablissement Technique at Angers, a French Army testing establishment, was given the task of testing practically all the light liaison vehicles in existence, among them the Cournil, Stemat, a vehicle constructed with the financial support of a subsidiary of the French coul industry, the American Motors M 151 A2 and CJ 5, the Toyota Land Cruiser, the Fiat 107 AD Nuova Campagnola, the Land Rover 88 with gasoline engine and a prototype called the Military Prototype, combining the engine and mechanical parts of the Range Rover with the bodywork of the Land Rover 109. All of them were tested systematically. In the course of these tests the contacts made with Citroen, Renault and Peugeot suggested a new approach which consisted in investigating whether existing vehicles might be "Frenchified". Such a programme would offer the advantage of obtaining vehicles which were already well proven and which had already been in production for some time in their country of origin, promising substantial economies in research and

Citroen C44 near La Courtine during the maneuvers in October 1979.

development without it being necessary to import units from abroad and thus exclude the French automobile industry from the programme.

Renault therefore investigated the "Frenchification" of Fiat's Nuova Campagnola with the gasoline engine of the R. 20 saloon, Peugeot chose the Mercedes-Benz G chassis into which they fitted the 504 engine, while Citroen turned to the VW Iltis which was fitted with the engine of the CX Athena. In July 1978, after an initial series of trials at Angers, the Peugeot P4 and the TR M 500 from Renault Vehicules Industriels were presented officially, The Citroen C 44 prototype followed soon afterwards.

During the following year 10 vehicles of each model were subjected to an experiment in accelerated ageing in the hands of units of the 15th Division and the 9th Marine division. The vehicle selected will replace the 8,014 Jeeps which were still in service in the French Army in 1980, that is :

3,514 Hotchkiss M 201
1,830 Willys MB
870 Ford GPW
280 Ford and Willys duel control driving instruction models.
270 Hotchkiss M 201 Milan
1,250 Hotchkiss M 201 with 106 SR or ENTAC

In addition, the French Air Force had 422 Hotchkiss, Ford or Willys vehicles.

Finally, most of the 7,064 Citroen Meharis in service (5,595 24 volt Meharis wired for radio, 778 not wired for radio, 691 12 volt with dual controls for driving instruction) will be replaced by the Citroen A 4 × 4, 5,000 of which are being ordered.

The order for the main Jeep replacement was placed with Peugeot in February 1981, covering 15,000 P 4 light all-terrain vehicles based on the G Series Mercedes-Benz to be delivered in several batches. The standard P 4 has a carrying capacity of 750 kg., is driven by the 505 petrol engine (87 b.h.p. at 5,000 r.p.m.) and has the 604 gearbox. It has a range of 600 km and a top speed of 110 km/h. The rear axle has a hydraulic differential lock which can be engaged on the move. A single lever engages the drive to the front axle and the reduction gear which can also be done on the move. The new VLTT will be equipped with a Milan anti-tank missile, a machine gun or a Rasura radar in the short chassis version with folding canvas top and a command post radio installation in the long chassis version with steel panelled body. Some of the vehicles may be equiped later on with the Peugeot XD 2,500 c.c. diesel engine.

Peugeot P 4 on test at the La Courtine camp during the October 1979 manœuvres.

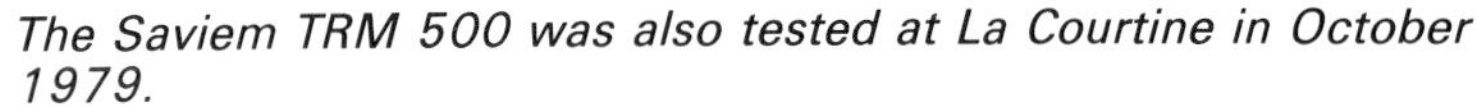

The Saviem TRM 500 was also tested at La Courtine in October 1979.

An illustration from a Willys catalogue showing a pre-production civilian Jeep based on a military MB with modifications planned for the future CJ 2 such as spare wheel mounting, rear tailgate, hood with side curtains and the name "Jeep" on bonnet and windscreen.

THE JEEP AND THE CIVILI

Having got itself adopted by the Armed Forces during the war, the Jeep embarked on a long civilian career immediately afterwards. But unlike other vehicles such as the Dodge or the GMC which could easily be adapted for civilian use, the Jeep suffered from two major handicaps; its carrying capacity was too small to make it a real utility vehicle and its typically military bodywork lacked platform space and any semblance of comfort.

The sole arguments in the Jeep's favour were its performance, its very reasonable price and the general shortage of motor vehicles throughout the world. Consequently many specialists tried either to provide a modest degree of comfort with a body better adapted to the needs of civilian drivers or to modify it fairly extensively to turn it into a small truck.

In fact, these possibilities had not passed unnoticed by Willys who started design work on a civilian version of the Jeep in 1945.

This was marketed under the name of CJ 2, then CJ 2A Universal and was a modified and improved MB (engine with gear-driven camshaft, fuel filler outside the body, drop-down tailgate, simpler top with side curtains, power take-off to drive machinery etc.).

In two years the catalogue of industrial adaptations approved by Willys became a bulky volume of more than 200 pages. Jeeps for use in agriculture had their own catalogue of approved equipment which was even bigger than the industrial one.

However, sales of this modernised version of the Jeep were practically confined to the American Continent, Europe and its colonies of those days relying upon war surplus for vehicles at reasonable prices. In France the government came into possession of a large quantity of motor vehicles as a result of the memorandum signed on May 28 1946 in Washington

From 1943 the American Government released some Jeeps to priority civilian users including farmers. Naturally the first to go were pre-production models like this Ford.

NS

Accident in the Place de l'Hotel de Ville, Paris, in 1948 caused by a garage proprietor's Jeep ignoring a red traffic light. It has a military number on the bonnet, a civilian number on the grille and surplus sale lot number on the side.

and the agreements in relation to the handing over of the property and surplus effects of the U.S. Army. Administration of the material acquired by the government was entrusted to the Société Nationale de Ventes de Surplus, the national company organising sales of surplus. The SNVS first had to undertake the onerous task of collecting the vehicles, compiling an inventory and then making an assessment of their condition and value.

Distribution was organised according to rules laid down by the Ministry of National Economy which controlled the SNVS, and established the following procedure in relation to motor vehicles; *"After the Minister of National Economy has decided on the percentages to be allocated to Metropolitan France, North Africa, to the Colonies and for export; the vehicles remaining in the parks may be sold"*. In

Jeep converted into a mobile library in Rome in 1947.

short it was the *"surplus"* of the Surplus which was available for sale!

Moreover it is necessary to make a distinction within this surplus of Surplus, between vehicles in running order ant those which were not, the regulation being different for the two categories.

This is how it was explained in July 1947 by Pierre Lenoir in the Revue Technique Automobile :

M.L. Philippe, a garage proprietor at Remiremont in the Vosges converted his Jeep into an articulated goods vehicle.

"The Jeep is the subject of special provisions. In the beginning about 20,000 of them were listed as in running order, of which 10,000 were immediately taken over by the Army. 3,250 were allocated to North Africa, 1,500 to the Colonies and 250 for export. There thus remained 6,968 Jeeps for other government departments and the civilian market."

Vehicles in this last category were allocated as follows. A "national quota" of 1,523 was shared out among government departments, the most important being Public Works, which received 600, while 450 went to the Agricultural Services. A further special quota of 150 was allocated to certain priority enterprises which had not been provided for at the level of the départements.

5,295 Jeeps have been distributed to the regions, and almost all have been allocated by the prefects through local motor trade distributors and dealers. In distributing this quota to local government authorities a proportion, varying according to the region, was reserved for agriculture and this absorbed a total of 1,557 units. Total allocations to the most favoured départements were as follows : Seine 484, Nord 207, Seine-et-Oise : 130, Bouches du Rhône : 126, etc.

Vehicles described as wrecks are the remains of Jeeps from which it is still possible to obtain spare

(continued p. 239)

Hotchkiss got SINPAR to design and build a winch for the Jeep but it increased the weight on the front axle and was soon removed. The vehicle, with its cut down grille, was used for communications work until 1970 by its parent Company.

SINPAR also produced this arc welding equipment driven by a take-off in the centre of the vehicle. Willys offered similar installations on the CJ 2 in the USA.

Verraro converted the Jeep for road-rail use for railways of one meter gauge. The central jack was used to turn the Jeep round or change wheels from rail to road. Eight were delivered to the army in Indo-China, one to the Malaga railways, one to the railway at Gsafa in Tunisia and two to the railways in Corsica. The Jeep loco could pull 80 tons of goods wagons but suffered greatly in the process, which is why few were built.

The SINPAR power take-off could drive a pulley at the rear of the Jeep to operate agricultural machinery.

Station wagon body with timber frame and steel panels built in England at the end of the fifties.

Agricultural contractors still use the Jeep for crop spraying and other treatment as it can move from farm to farm faster than a tractor (Brittany 1980).

Jeep used by the Esso agricultural department with trailer for spraying fruit trees. Most Jeeps used for this work carried a tank at the rear, a high pressure pump in place of the passenger and two spray units on folding arms.

In 1951 Jeeps used for geological surveys round Parentis in the Arcachon Basin were fitted with special exhausts to minimise the risk of forest fires.

Before starting production of their hydraulic excavators in 1951, Poclain made agricultural equipment. Their multi-purpose trailer, the Trirou enabled the Jeep to tow much more than the 1/4 ton Trailer originally intended (1951).

A Poclain post carrier of 1952 towed by a Jeep which seems to handle it with no great difficulty.

Lambert loader produced by Poclain for Jeep or Dodge from 1951.
Jean-Louis Duval-Arnoult, engineer, electrician, mechanic and aeronautical engineer joined the company in 1947 and devised a lightweight loader on a Dodge 4×4 attached to the bumper by two bolts. It was equipped with a hay fork and a dung bucket. This was the first machine to use the single-acting hydraulic jack later employed on the hydraulic shovel.
This loader could lift 500 kg. to 2.60 M.

Vehicles for the 1950 Tour de France on show in the Place de la Concorde in Paris. The white Jeeps carried on the bonnet the national colours of the teams to which they were allocated.

In the Pyrenees in the same year a Jeep assists a competitor. It has road lighting from Citroën Traction Avant and road tyres.

THE JEEP AND THE TOUR DE FRANCE

When the Tour de France was restarted in 1947 one of the problems the company organising it had to solve was that of the vehicles to be used to accompany the Tour. Until then standard cars, usually convertibles, had been used. They were not without their uses but they did not have all the qualities required.

When the technical team of the organisation tackled the problem, they found in the Jeep a vehicle which perfectly met their requirements, a convertible which cost little to buy and maintain, simple, with a body which lent itself easily to the mounting of racks to carry bicycles. In addition the mechanics or the Tour physician could easily lean out of the body to bring help to men or machines. For the 1947 Tour the organisers allocated one Jeep to the manager of each national racing team. They were painted

Racks carrying bicycles and spare wheels on a Jeep in the Tour de France 1953.

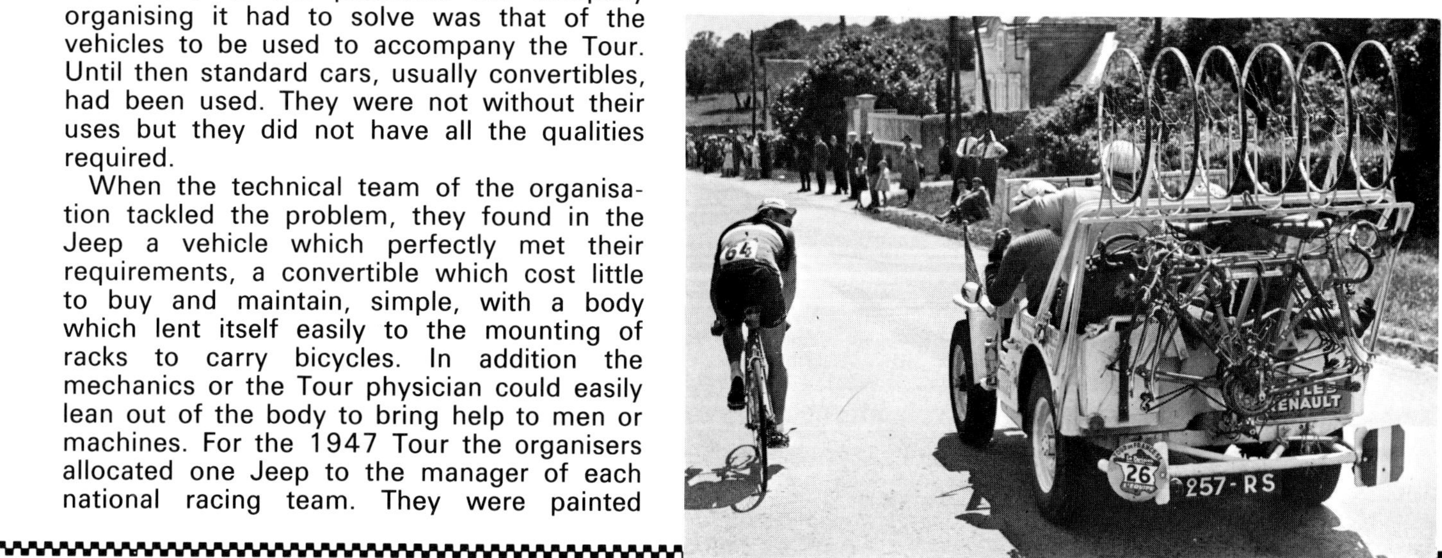

white, the colour of the Tour vehicles with the flag of the national team to which they were attached in colour on the front. Up to 1955 about 18 Jeeps were used by the Tour de France. Beside the bicycle racks, these Jeeps were fitted with more powerful headlamps from front-wheel drive Citroens and tyres with highway tread patterns which were more comfortable and improved the road holding.

Film cameramen and press photographers also adopted the Jeep as it allowed them to work on a much more stable platform than a motor cycle and one free of all obstructions.

In 1952 the reporters of Paris Match were the first to follow the Tour with a Delahaye instead of an American Jeep. When the time came to replace the Jeep, Delahaye devised a Tour de France version of the VLR-D but it was not bought by the Tour organisers who turned instead to the Peugeot 203 with the doors either removed or cut down. Once again, Delahaye had thought of something too splendid, too expensive and too sophisticated.

The Italian national team took part in at least one Tour, that of 1952, using Alfa Romeo Mattas.

Camermen and press photographers found the Jeep the ideal vehicle for covering the Tour de France (1952).

parts in more or less good condition. It should be made clear right away that buyers of Jeeps for reconditioning will be given facilities for acquiring wrecks. The Army has already reserved a fairly large number of Jeeps for reconditioning and little more than a thousand Jeeps in this category are left for the private sector in Metropolitan France. Of this number, 500 are reserved for agriculture and will be allocated through the Union Centrale des Coopératives Agricoles... In principle, the names of the recipients have already been decided. There are therefore only another 500 left to allocate and they are all reserved for priority buyers.

However, it is not impossible that with the reduction in military funds, the Army may decide not to take up the whole or part of its quota of Jeeps for reconditioning, which may offer some hope to those who still hold priority certificates waiting to be honoured.

Those who are not totally committed to the Jeep as a light truck may possibly be able to settle for one of the 3/4-ton Dodges which are being disposed of without restriction.

The instruction book translated into French by the SNVS, had this to say about the Jeep :

"Although conceived for military purposes, the Jeep has numerous civilian applications, notably in the transport of people and goods in difficult country (public works building sites, agriculture, colonial territories, etc.) Given the shortage of communications vehicles it can also prove an invaluable emergency ambulance.

Further, because of its technical characteristics, (reserve of power, numerous gear ratios, four-wheel drive) the Jeep can be converted and adapted for special requirements :

- *General purpose vehicles for agriculture (power take-off, governor).*
- *Small garage breakdown vehicles.*
- *Fire fighting vehicles (motor-driven pump).*

Moreover, in view of the general state of the French economy, the carburetter manufacturers have designed special equipment which makes it possible to reduce the consumption by 10 to 20 % according to the way the vehicle is used".

Having absorbed the contents of this manual, the happy Jeep owner usually embarked on a programme of improvement. For this he relied either on specialist constructors or on his own efforts.

Among those supplying new bodies, LP coachworks were the first to attract attention by

(continued p. 242)

Jeep in use for reporting floods in the Paris region in 1956 by French television which employs several MBs and CJ 2A vehicles.

At the end of the fifties, the French police used Jeeps rebuilt and converted by the Central Workshops at Limoges. Some were tried out as light radio vehicles like this one seen near Vincennes outside Paris.

Kam, an elephant from the Bertram Mills Circus drives his own specially adapted Jeep in the streets of London at Christmas, 1955.

An advertisement by Berthet of Rives in the Alps for a Jeep snow plough.

CHASSE-NEIGE POLYVALENT pour JEEP

transformable en étrave et lame biaise orientable

TRACTEURS
DODGE, ETC.

BREVETÉ
LICENCE
SOFIA-BALLU

MÉDAILLE
DE BRONZE
CONCOURS 1954

Matériel d'intervention rapide à grand rendement

TRIANGLES REMORQUES - ETRAVES - LAMES BIAISES
SABL[illegible]SES ETC

A Brittany fish merchant's Jeep recalling in its general outline the half-ton 4×4 prototype built by Willys.

The Carrosserie Arnault at Garches built this wooden station wagon body which the French called a "canadienne" for one of their clients. Similar bodies were built on Ford-Poissy and Simca 8 chassis.

The Baby Duriez.

creating a station wagon body for the Jeep with timber frame and wood panelling. But the leader in this sphere was, without any doubt, Duriez.

Like many other constructors, Duriez produced a Jeep with station wagon bodywork which he called the "Baby Duriez", but he became known mainly for his light trucks. His great idea was to provide more load space by increasing both the length and the width and moving the driving position forward.

At the same time, the payload could be increased by fitting reinforced suspension. Duriez offered kits containing all the parts needed to convert the military Jeep into a general purpose vehicle suitable for both agriculture and industry. In his catalogue he listed seven types of conversion kit.

Models	Code	Kerb Weight			Dimensions	Volume in m³
		Front	Rear	Total	L W H	
1) Cabin alone	CAB	820	300	1 120		
2) Cabin and rear frame	CSOAR	830	330	1 160		
3) Cabin with rear platform	CNP	840	380	1 220	2 500 × 1 750	
4) Cabin, platform and sides	CPR	845	440	1 285	2 490 × 1 700	
5) Cabin, platform, sides, tilt	CPRB	850	485	1 335	2 490 × 1 700 × 1 600	6 à 6,5
6) Baker's van	BOUL	840	570	1 410	2 365 × 1 500 × 1 400	4,5
7) Station wagon	CAN	855	615	1 470	2 365 × 1 500 × 1 400	4,5

Light truck with forward control by Duriez who also built a station wagon and a steel-panelled van on the Jeep chassis.

One of the first Duriez advertisements in the Revue Technique Automobile.

Generally speaking, the conversion went like this. For a start the Jeep had to be stripped down to the bare chassis and a number of components then had to be altered :

- Cut and bend the gear lever;
- Cut back the bracket securing the cross member supporting the gearbox on the right side of the chassis to make space for the fuel tank;
- Cut the rivets holding the battery mounting and the front body mounting;
- Cut a notch in the front bumper to allow clearance for the track rod;
- Lenghthen the drag link and modify the drop arm.

Having completed these modifications one could start to build up the Duriez. Side member extensions were attached to the chassis by U bolts and through bolts to increase the overhang at front and rear. Cross members were already welded to them to extend the lateral support for the bodywork.

The Duriez had a forward control cabin but the engine remained in its original position, so the kit provided for re-positioning of some items, and remote control of others. As the radiator was moved forward, longer hoses were provided and the fan had to be moved forward too, by means of a shaft fixed to the pulley driving the water pump. As the original scuttle no longer existed, the parking brake was mounted vertically on the floor and extended linkages allowed the accelerator, brake and clutch pedals to be moved forward. The steering column was attached to a new support to bring it more upright. Gearbox and transfer box were provided with rather rustic systems of remote control.

Once the chassis had been modified in this way it only remained to mount on it the chosen body, complete the interior trim and install the electric wiring. According to Duriez, all these operations could be undertaken by a good amateur mechanic or a garage man without any great difficulty.

In England too, many conversions emerged, notably from Burleigh, Farmcraft, FWD Motors, Wicks Autos, Hall coachworks and above all, Metamet.

Metamet (Metal and Metal) had specialised in the manufacture of components for automatic weapons during the war, but from 1946 they entered the civilian market selling standard Jeeps which they bought from the Canadian and American forces.

The Polish-born director of the company, Mr. S. Chrulew, used a Jeep for all his personal motoring and following the example of some of the GIs, he decided to civilise the car by giving it fixed half doors and a folding top, offering either partial or total protection against the elements. Removable rhodoid side windows were mounted on the half doors.

This adaptation attracted so much attention on his travels that he decided to put it into production and sell it under the name *"5 in 1"* indicating the five degrees of protection which it offered.

- vehicle open;
- with side windows erected;
- top erected without side windows;
- top erected only over the rear seats and side windows installed;
- completely closed with full top erected and side windows.

Other variants on the same theme soon followed; the *"Farmer"* with drop-down rear tailgate and higher top, the station wagon in wood with lengthened rear overhang and a *"5 in 1"* with metal hard top.

These models had a certain success and nearly 200 were built up to 1960, but more original designs followed. Metamet became specialists in lengthening Jeep chassis, increasing the wheelbase by 50 cm., which allowed it to be fitted with much more practical bodywork to create genuine light trucks or vans, to say nothing of the de luxe station wagon. The chassis was lengthened quite simply by cutting it in two just behind the transfer gearbox and welding in inverted U section extensions, while the rear transmission was lengthened by use of salvaged tubing.

Other projects such as an articulated Jeep were considered but never brought to fruition.

Among other conversions carried out around the world, there was the small omnibus built by Beutler in Switzerland. Italy deserves mention for the articulated passenger vehicle built by Orlandi using a Jeep as the tractor unit; this very elegant conversion was capable of carrying 20 passengers. Sometimes the Jeep was re-bodied and made to look like an American saloon with wide grille and tail fins and there were other things even worse.

However, the first prize for these conversions must go to the Philippines. In the immediate post-war period, when a lot of firms which had begun special-

John Burleigh (Automobiles) Ltd. of London built this station wagon with timber frame and steel panels on a Jeep chassis lengthened by 51 cm. There were two doors on the left but only one on the right.

The "Five-in-One" built privately in 1947 led to the launching of the Metamet Jeep conversions on the British market.

The Metamet station wagon on a Jeep chassis with wheelbase lengthened to 2.54 M resembled that built by Willys in the USA. This was the most luxurious Metamet model.

ising in the re-sale of Jeeps were competing against each other to get vehicles in the best possible condition, so that they would not need much work on them, one business in the Philippines was ready to buy Jeeps in any condition so long as they were as cheap as possible. As they hsd no direct competitors, hundreds of Jeeps arrived by the boatload in the Philippines, sent on their way by derisive Europeans who saw an opportunity to get good prices for second-rate material.

On arrival in the Philippines, the Jeeps were completely stripped down and all mechanical parts were restored to new condition for use in the construction of *"Jeepneys"*.

The mechanical elements of the Jeep were installed in a chassis which had been considerably lengthened and was generally fitted with bodywork capable of carrying nearly ten people under an awning which served solely to shade them from the sun. The bodywork was decorated by hand with florid arabesques in bright colours set off by chromium plating in American style. Such extensive work on the body and its decoration was only possible because of the ridiculously low cost of local labour. So these Jeepneys, which cost very little to make, became practically the national motor vehicle. Even now, the Philippines are the country with the highest concentration of Jeeps, or rather of Jeepneys.

Apart from these coachwork conversions, the Jeep underwent many mechanical modifications, particularly in the agricultural field. In 1946 agricultural production had hit rock bottom, farms and estates having been ruined by the war. There was a shortage of horses and oxen to pull implements or farm carts

Metamet "10cwt truck" with wooden body on a chassis with extended rear overhang.

and labour was short too, so farming needed to be mechanised. But there were too few tractors and the Marshall Plan was far from providing all that was needed.

And so, in all parts of France, handymen started up in business getting jeeps from the Surplus and modifying them for use as tractors. Among them was Bernard Cournil, at Aurillac, who modified nearly 600 Jeeps for the farmers of the Cantal and the surrounding regions before creating his own all-terrain tractor.

The Poclain workshops produced a hay loader under Lambert licence, Esso-Standard had a department using a number of specialised machines, some buyers mounted Guinard pumps on Jeeps and there were many other conversions, often done on their own account by farmers, the eternal jacks of all trades.

The essential mechanical features of these transformations were the fitting of a power take-off or mechanical lifting gear at the rear, a strengthened clutch and an *"agricultural"* gear with a ratio of 2.77 or 2.43 to 1 which gave the extra-low speed necessary for pulling a plough.

Obviously the heavy fuel consumption of the Jeep was a handicap. Various more economical carburetters were produced such as the Solex type 32 AIC without accelerator pump, 32 AIC downdraught with accelerator pump, 30 RFAI downdraught with governor but no accelerator pump and the Zenith type EX downdraught, but the real solution arrived at the end of the nineteen fifties with the fitting of light diesel engines, mostly the Peugeot Indenor, with some from Perkins and Mercedes-Benz.

In 1948 Orlandi built a number of these semi-trailers for the Jeep, some carrying up to 20 passengers, others to carry goods.

On the same chassis as their station wagon Metamet offered the Caravaner, what is now known as a recreation vehicle. It had a bench-type front seat for three.

Spapa, a company laying asphalt paving was still using a Jeep in 1979 to carry sealing materials for surfaces in parking garages and similar places where its low height and four-wheel drive made it ideal.

An unusual conversion into a half-track used by the Gertraud in Die Sonne hotel at Salzburg in Austria in 1959.

Vehicle used by an Austrian mountain fire brigade which passed into the hands of a private owner in 1967.

Not long ago Barreiros in Spain were planning to bring out a conversion kit for the MB with their own diesel engine. Although this conversion was devised for military purposes, it would be possible to use it on civilian vehicles. The object of the scheme was to equip Jeeps with diesel engines in time of war, but to leave them with their petrol engines as an *"economy measure"* in peace time.

Austria, with its mountainous farmland, was and is an ideal terrain for the Jeep, where it is used in many different forms, some of which have only a remote relationship with the vehicle from which they are derived. There too the diesel has become popular.

Among the first civilian uses for the Jeep which Willys investigated was its employment as a fire-fighting vehicle. The first prototype, based on an MB with

(Continued on p. 252)

"Note by SOFIA on the Conversion of the Military Willys Jeep for Use in Haulage or Agriculture"

1) The reduction gear.

Use of military Jeeps from Allied war surplus for heavy haulage or agricultural work makes it necessary to obtain a higher torque at the outlet from the transfer box or at the wheels of the vehicle and lower intermediate gears.

Thanks to a minor mechanical modification the SOFIA reduction gear raises the torque available at the output end of the transfer box by more than 60 % compared with that of the military Jeep.

Heavy haulage or agricultural tasks can be carried out in the intermediate gears, achieving the maximum traction and driving all four wheels by engaging the drive to the front axle.

The SOFIA reduction gear reduces the intermediate gears by a ratio of 3 : 5 in comparison with those of the military Jeep while retaining the road speeds attainable with the transfer box disconnected.

The capabilities of the military Jeep are thus appreciably improved both in heavy haulage on roads and over varied country and also in work which employs the Jeep's power take off. Speeds obtained at 1,000 engine r.p.m. are

1st gear 3.5 km/h instead of 6 km/h
2nd gear 6.8 km/h
3rd gear 10.8 km/h

2) Power Take-off

The power take-off designed by SOFIA consists of

a) central take-off shaft
b) drive shaft with universal joint
c) two-speed angle drive
d) support bearing
e) pulley of 200 mm. diameter

The central shaft provides a drive for all equipment mounted on the platform of the Jeep such as pumps, generators, compressors, welding equipement etc. as well as a side mounted grass cutter. The complete assembly either with two-speed angle drive or with support bearings can operate all agricultural machinery either towed or carried, harvesters, reapers, borers, fertiliser spreaders, wheeled crop sprayers, etc.

The pulley can drive all fixed agricultural machinery; threshing machine, saw, pump, straw cutter, binder, crushing mill etc.

The r.p.m. obtained at the outlet from the transfer box are as follows :

Engine r.p.m.	1st gear	2nd gear	3rd gear	Reverse gear
1 000	375	639	1 000	281
1 500	562	958	1 500	421
2 000	750	1 278	2 000	562
2 500	937	1 597	2 500	702

An Austrian Jeep with lengthened chassis fitted with a snow plough in 1967.

Another former fire service vehicle taken over by a private owner in Austria in 1970.

A breakdown vehicle seen in Austria in 1972 with a cab from a Volkswagen Transporter.

And who would recognise a Jeep under this camouflage seen in Austria in 1972?

A typical Austrian farm vehicle seen in 1972 with a lenghened chassis and diesel engine.

A Jeepney in Manila in 1979. Somewhere underneath this baroque decoration are the remains of a Jeep.

THE RALLYE DES CIMES

The Jeep made its French debut as an all-terrain recreation vehicle in the Basque country. At the end of the war, the Basques discovered the vehicle which they knew little about up to then and quickly took it up as working equipment to replace the horse, the mule, even their canvas shoes. Better than any animal or tractor, the Jeep seemed the ideal vehicle for all sorts of journeys on routes which as Jacques Loste wrote in *"L'Argus"* at the time, *"went from macadam roads to mule tracks by way of loamy lanes under the trees, meadows with wet top soil, footpaths overgrown with bracken and stretches of small stones like ballast for railway lines, all garnished with rocks and jagged outcrops lying in wait for the tyres. And the picture would still be incomplete if one did not add that these tracks oppose the passage of the cars with gradients that it would be difficult to climb on foot."*

Hence in collecting the milk or the cheese the drivers needed a certain expertise to exploit the possibilities of the Jeep to the best advantage.

The Basque temperament led the inhabitants of Tardets, Licq Atherey and Mauleon to compete against each other in driving skill and endurance on their Jeeps. From bets and challenges, local champions sprang up everywhere and many fierce races resulted. Eventually this encouraged the Mayor of Licq Atherey to organise an official event on a course from Licq to the Coll d'Orgambisceda and back. On September 9 1951 Sauveur Bouchet presided over the first race for Jeeps in France under somewhat vague regulations which indicate the spirit in which the event was organised.

Art. 1. The event is open to all holders of a driving licence, amateur, professional, civil and military.

Art. 2. Vehicles must be Jeeps or similar.

Art. 3. Civilian drivers must arrange an extension of their car insurance for the duration of the race.

Art. 4. The Committee reserves the right to eliminate any competitor not showing the necessary skill. The poor state of a vehicle may be taken into account.

Art. 5. Competitors who are eliminated before the start may not make any protest or demand any compensation.

Art. 6. Starting order will be by draw one hour before the start. In principle the first vehicle will

Start of the first Jeep race at Licq-Atherey, September 9 1951, which was won by Sauveur Bouchet. It was a tough event; of 20 entries only eight started and three finished.

Second "Rallye des Cimes". It was banned by the Automobile Club de France but was run after high-level intervention.

The Rallye des Cimes becomes safety conscious. Crash helmets and roll arch become compulsory but the ruts remain the same.

start at 13 hours, the others departing at 10-minute intervals.

There were 20 entrants but as a result of accidents during practice, only eight competitors started, six reached the Col and only three crossed the finishing line. For the French Army which had entered four Jeeps from the 5th District of Toulouse it was a total defeat. Of the two Delahayes which took part, one had an accident on the road and the other disappeared into a ravine during the descent, fortunately without its driver. The result of this first race was a victory for Sauveur Bouchet followed by two drivers from Tarbes but generally speaking, as the results proved, it was a severe test of the car, in fact too severe.

The following year the course was lengthened as far as St.Jean Pied-de-Port. The army did not take part and stayed away until 1969. Ten competitors took part in this event which was forbidden by the Automobile Club de France and only took place after the intervention of prominent personalities who well knew how important the event could be for the Basque country and even for the automobile industry. This time Lapieza was the best of the seven drivers to reach the finish.

In 1953, the third edition of the *"Circuit des Cimes"* established the success of the event. The public arrived in large numbers, the press was enthusiastic and the event was authorised by the Federation du Sport Automobile and run under the patronage of the Automobile Club Basco-Bearnais. The route was the same as in 1951 with four laps of a 2 km. circuit round Licq in addition.

Of nine vehicles entered four Jeeps finished the course together with a VDB amphibian. This was

Starting in 1969 the French Army made serious efforts in the Rallye des Cimes with parachute troops of the first regiment of Hussars. They set up a communications system to transmit times of all competitors.

The world fame of the Rallye des Cimes attracted crowds to the most difficult sections like this dried up river bed. Despite the presence of more recent and more sophisticated cross-country vehicles the Jeep remains supreme.

the Vehicule du Bled, the little vehicle designed by Georges Irat and powered by an engine of only 5 CV. The winner was Guillaume Bouchet, brother of Sauveur.

In 1955 the tragedy of the Le Mans 24 hours affected all motoring events. The authorities banned the rally but by now its reputation had spread beyond the frontiers. Already a similar event had been organised in Chile and other countries like the United States, Italy and Rumania were thinking about it.

The Rallye des Cimes was not restarted until 1960. This fourth edition began with a detour to the Cemetery at Licq where all the competitors gathered at the grave of Sauveur Bouchet who had died two years earlier. This time the regulations were much more comprehensive, the vehicles were equipped with roll bars and the drivers had to wear crash helmets.

The number of makes involved had increased, testifying tot the interest that manufacturers were now taking in the event. They included Renault-Sinpar, Citroen, BMC, DKW, Kaiser, Hotchkiss, Fiat and Frua. But the Jeep continued to reign supreme. In 1960 Marcel Ricarte won the event; in 1961 it was Arnaud Bouchet, the son of Sauveur, in 1962 Jean Iriart won and 1963 Joseph Etchecopar. All were on Willys or Ford Jeeps.

Over the years the Rallye des Cimes became a classic. From 1965 it was supported by Mobil, then from 1969 by the Societe Nationale des Petroles d'Aquitaine. In 1969 also the Army came back with paratroopers of the 1st Regiment of Hussars. The involvement of the Army was complete both at the level of organisation, with its communications network and command post and in the quality of the entries. But for all that it was not until 1971 that the Army's efforts were rewarded with a victory by Lt. de Lantivy who naturally drove a Willys Jeep. Since then the Federation Française du Sport Automobile has officially recognised this type of race for special vehicles equipped for cross country work and in 1972 it set up an all-terrain event committee, with Andre Olibet as chairman. At that time there were ten events listed. In 1977 there were 24 but with its 100 entrants and its formidable difficulties the Rallye des Cimes remains the premier event.

Since then other events have emerged such as the Rallye Infernal or more recently the Paris-Dakar rally and the 5 X 5. But for everyone Sauveur Bouchet's rally remains something unique. It has become the equivalent of the Le Mans 24 Hours in cross country racing.

In 1965 the ninth Rallye des Cimes was a successful event organised by the Automobile Club Basco-Béarnais with the support of Mobil who contributed prizes worthy of the event. (Left) a Hotchkiss Jeep enters a village by a mule track.

Proud of having won the 1971 Rallye des Cimes with Lt. De Lantivy, the French Army exhibited his Jeep at the Salon de l'Automobile.

An MB which was the prototype for the future fire fighting version of the Jeep CJ 2.

a water pump in front, began its tests in USA in 1946 and led to production models on CJ 2 and CJ 2A. During the hostilities some Jeeps had already been used as light fire-fighting vehicles, fitted with CO2 and water extinguishers which were mounted on the wings, in place of the rear seat and in the jerrican holder, while an extensible ladder was carried on the outside of the body.

More recently the Jeep has been employed in the USA as a light mobile pump for country use, the equipment and the water tank being carried on a two-wheeled trailer. It is also used as an emergency vehicle for the American forest fire services carrying fire-fighting equipment with a high pressure water pump at front and a plough at the rear with lifting gear, to create improvised fire breaks.

Incidentally, a engthened Bantam, equipped with extinguishers and bottles of CO2 was used by the factory for its own fire protection during the war.

In France, the Jeep first came into use by the fire

The Guinard Jeep as built for the Landes of Gascony. Later they were given a vertical exhaust for the motor pump and an extra hose reel at the rear. Jeeps of the same model were delivered to other fire units including Marseilles. The motor pump with Train engine takes the place of the front passenger.

Rear view of the same vehicle showing the agricultural tractor seat at the rear which was uncomfortable and could be dangerous.

services when the forest fire service of the Landes of Gascony was re-equipped with 157 Guinard engines. On the creation of the corps in 1947, 65 of the Jeeps were placed at the disposal of the Gironde, five of which were later handed over to the Lot-et-Garonne. These vehicles were equipped with a motor-driven Guinard type GMOA — 6 six-stage centrifugal pump. Because of its shape, this pump became known as *"the sausage"*. It was driven by a 3 b.h.p. Train engine and had a nominal output of 100 litres/minute at 5 bars.

As the pump was installed in place of the front passenger, the second member of the crew perched precariously on an iron tractor seat high up behing the driver where he did nothing to improve the weight distribution of the vehicle or its centre of gravity. As the suspension had not been strengthened, the Jeep could only carry 250 litres of water in a little oval tank.

These Jeeps were the first fire engines ordered for the Landes forest, the Half-Track and the GMC only arriving in the following years.

From 1946 to 1965 Guinard fitted out 160 Jeeps and supplied 28 hydraulic units consisting of pump and water tank for buyers to mount on their own vehicles. In 1964 a local dealer bought the last of these kits to mount on a Jeep destined for the Haute-Vienne département. Next to the Landes of Gascony, the biggest user of Jeeps was the Bouches du Rhone département, with 25 engines in 1950 and 1951. In Lyons, Maheu-Labrosse built a prototype of a light fire pump unit on a Jeep base, with a rear-mounted pump, designed to meet the needs of rural fire brigades, but nothing came of it.

The cicil Defence authority, then known as the Service Nationale de la Protection Civile, ordered three light tanker trucks from Maheu-Labrosse with high pressure motor driven pumps, two in 1958 and one in 1960. These were the only fire engines to be built from scratch on the Hotchkiss M 201 Jeep.

In October 1951 technical document MI — DOC No.12 first prescribed the equipment required and the tank capacity for water-carrying fire engines and under the heading *"Light water tender for forest fires"* it said *"chassis Jeep type or similar"*. *"Similar"* in this context could only mean the Delahaye. This really gave the Jeep a walkover because the VLR-D was effectively ruled out by its price and various problems of adaptation.

Hence for firemen the Jeep was the synonym for a light tanker fire engine for use against forest right up

In 1958 Maheu-Labrosse built three tanker fire tenders on M 201 chassis for the Civil Defense authorities for use against forest fires. To conserve the water in the small 250-litre tank the vehicle had high pressure equipment with a low output of 33 l/min. at 40 bars. Each reel carried 50 m of hose.

Guinard Jeeps were used by the CRS in 1951 against the great Maures fires. This one has a stretcher carrier.

For many years Paris Airport used Jeeps with cylinders of CO2 for fire duties at parking areas and secondary airfields. The fabric awning seen at Saint-Cyr in 1970 protects the gas cylinders from the sun.

Besides its role in fighting forest fires the Jeep is used as a light rural fire engine like this at the Lake of Annecy in 1971, carrying hoses, a ladder, two or three firemen and towing a motor pump.

to 1970, when a new official specification raised the tank capacity to 600 litres, which was more than the Jeep could handle.

From 1956 onwards practically all of these light fire engines were built by Guinard or Maheu-Labrosse on the JH 102 and some, more rarely, on the HWL, CJ 5 or CJ 6. Of course some local constructors and a considerable humber of fire brigades built their own.

Most of the MB and GPW models which had been converted into light forest fire engines had their equipment removed at the end of the 1960s and were then used as communications vehicles. Apart from the small quantity of water they carried, these converted Jeeps had come to be regarded as too heavy, with too high a centre of gravity, which made them dangeroux to drive both on the. road and across country.

For this reason Hotchkiss fitted the JH 102 with a special reinforced suspension with shock absorbers by

Jeeps were often used to carry 250 kg powder extinguishers (Type 3000 B) capable of extinguishing a fire of 3,000 litres of gasoline. Sicli provided some for protection of the Monaco race circuit in 1960. This unit has two Biro 250 kg powder extinguishers.

The French Air Ministry ordered a number of Jeeps from SIDES with gas cylinders mainly for use on aerodromes with light traffic.

Guinard. No Willys or Ford Jeep was ever given official approval by the technical services of the Ministry of the Interior. When official trials were organised in 1955 both Guinard and Maheu-Labrosse preferred to enter early examples of the JH 102 for which they had high hopes.

Generally speaking these hopes were not realised. Guinard, for example only built 98 light tanker forest fire engines conforming with the official specification between 1956 and 1964. Sides, Biro and Sicli adapted the MB or the GPW as first aid fire engines for aerodromes, usually with powder extinguishers or 250 kg of Tribromofluor, a gas extinguisher.

The most ambitious conversions were undoubtedly those carried out in Austria for fire brigades in the mounttains. Although they were generally withdrawn from service nearly ten years ago, and replaced by new fire engines, MB or GPW Jeeps still turn up almost unregocnisable, here and there in Austria.

Light fire engine with American pump in Belgium at the end of the fifties. The first prototype built by Maheu-Labrosse in France was very similar.

A light truck on a lengthened Jeep chassis used by the Furth-Triesting volunteer mountain fire brigade in Austria in August 1962.

The volunteer fire brigade at Langenwang in Austria used two light Jeep motor pumps seen here in 1964. One pump in front was protected by a tarpaulin, the other was on a trailer. Hoses were carried in a rack at the rear and suction pipes on the hood frame.

A Jeep with lengthened chassis converted into a truck to carry a foam generator at Knittelfeld Austria in 1971. Nothing much remains to identify it as a Jeep...

Fire engine with Rosenbauer motor pump of 1,250 lit/minute delivery at Saint-Martin in Austria, 1964.

◀ *Light mountain fire engine with Rosenbauer pump in front on a lengthened MB chassis at Saalfelden, Austria, in 1969. It had four doors and comprehensive equipment. Most of them were eventually replaced by vehicles based on Ford Transit, Land Rover, Toyota or CJ 20.*

Something rare in Austria, A fire pump unit with the pump at the rear (Reinthal 1972).

Low carrying capacity meant that the MB and GPW were mainly used for communications like this one with loudspeaker.

A lengthened Jeep for personnel transport popular with Austrian fire brigades. Compare it with those used by the American Coast Guards.

GAZ 67 B seen in July 1954 at Trun-Gia in Indo-China at a meeting of French and Viet-Minh delegations discussing a cease-fire. It differed from the 67 and 67 A in new frontal design and an extra fuel tank under the front seats. The GAZ plant (Zavod Ineni Molotov, Gorki) built the MI in 1941 then the GAZ 64 and in 1942 the GAZ 67. Cross-country types like the 67 B owed mechanical inspiration to Ford, Willys and Bantam, engines being derived from Ford trucks built by the plant in the twenties while other parts showed Willys and Bantam influence.

THE JEEP CULT

In this chapter, which is not meant to be exhaustive we present some of the progeny of the Jeep, or at least some of the descendants which resemble it most, among them those like the Land Cruiser and the Land Rover which have vigorously challenged the success of their elder relative.

Nearly all the vehicle descended from it have been strongly influenced by its appearance, because the original bodywork was a perfect shape and a perfect design for the job. This is why it is fair to speak of a Jeep cult. And of course the Jeep itself continues its career in the shape of the current CJ 5,6 and 7, Golden Eagle or other Renegades, in spite of the industrial vicissitudes which were part of its post-war history.

But that is another story.

When the Jeep used on his farm began to wear out, Maurice Wilks, a Rover director, decided to produce a British vehicle to replace it. Gordon Bashford made the first drawings. This first Land Rover prototype was inevitably inspired by the Jeep because it used many of the parts including chassis, suspension, transfer box and axles. Engine, gearbox and central steering were Rover components. When the Land Rover Series I emerged in 1948 it was an all British and much improved design.

After testing a prototype in 1950, Alfa Romeo put the 1900 M called Matta (The Mad One) with independent front suspension into production. 2,000 were built for the Italian army and public services and a further 50 for civilian users.

The Fiat Campagnola made its debut at the Levant Fair in Bari in 1951 and from December 1951 to January 1952 it did the run from Algiers to Cape Town and return in 11 days 4 hours 54 minutes setting a new record.
In 1953 a diesel engine was offered. In 1955 the petrol engine went from 53 to 63 b.h.p. Electrical equipment was modified in 1959. On version A (1955) and the B series an oil radiator was fitted alongside the water radiator. Thousands were built by Fiat in Italy and by Zastava under licence in Yugoslavia. Production ceased in 1972.

The Mighty Mite was designed in 1950 by Ken F. Gregory as a light vehicle which could be carried by the helicopters of the period. Weighing 678 kg., practically the amount specified for the original Jeep, it could be slung under any helicopter used by the Marines, something then impossible for a normal Jeep. Built by Mid American Research Corporation at Wheatland, Pennsylvania, it had a Porsche flat-four engine, three-speed gearbox with step-down gear and permanent four-wheel drive. It could run with great ease on three wheels. The marines tested twelve prototypes up to 1954. This one is entering the prototype Sikorski SS 56 cargo helicopter watched by Col. Shepherd, Commandant of the US Marine Corps in January 1954.

Probably the best known of all the light Russian cross-country vehicles, the GAZ 69 was built from 1953 in the Gorki plant and later at Ulianovsk where it received the designation UAZ 69 AM. It was also built by ARO in Rumania as the M 461 in China. There was also a two-door version. Tens of thousands were built and exported during 20 years of production.

In 1951-52 DAF built two prototypes of the YAO 54. Contrary to custom, the front wheels, not the rear, drove it on hard roads. It had a central differential and torsion bar suspension. The engine was a petrol driven Hercules giving 60 b.h.p.

Developed from the Nuffield vehicle designed by Sir Alec Issigonis at the end of World War II, the Austin Champ emerged from the Wolseley factory in 1949 as the FV 1800 with new bodywork and a Rolls-Royce engine. The production model, the FV 1801 (A) was delivered from 1952 and 13,750 were built at Cofton Hackett up to 1955. The Champ was extremely complicated and never very successful. Some were built for the civilian market like the one illustrated with the Austin A 90 2,400 c.c. engine. They lacked the Snorkel fitted on the right of the bonnet of the military version. The history of the Champ is strangely like that of the Delahaye.

When the West German Army was re-established it too organised a competition for a light liaison vehicle and Goliath, Porsche and Auto Union entered. The Goliath factory at Bremen produced a vehicle with independent suspension and a two-cylinder two-stroke water-cooled engine of 950 c.c., Model 31. Fifty units were ordered by the German authorities and an improved version, Model 34 called the Jagdwagen was produced with a four-stroke engine. Production was soon terminated as the Auto Union was chosen for large scale production. The Goliath above is seen during maneuvers in 1954.

Having presented the prototype Model 597 in 1954, Porsche continued to develop it until 1958 although the German authorities had adopted the Auto Union. Like all Porsches up to then it had its 50 b.h.p. engine at the rear but failed to repeat the success of the Volkswagen Kubelwagen.

One of the most successful post-war cross-country vehicles is the Land Rover, still in production, which is worth a book to itself. These two with the short 86 in. wheelbase which replaced the 80 in. of the first series were lent to a Trans-Africa expedition by Oxford and Cambridge undergraduates via Paris, Algiers, Cairo, Addis Ababa and Nairobi in June 1954.

In the early fifties Toyota produced an all-terrain vehicle much influenced by the Bantam, as the manufacturer had built five prototypes based on a captured Bantam for the Japanese army during the war. The 85 b.h.p. engine was similar to a Chevrolet design. Nissan also produced their own Patrol which also showed Jeep influence.

◀ *Auto Union finally won the order from the Bundeswehr with the Munga. Power of the 3-cylinder two-stroke engine was raised from 38 to 44 b.h.p. during its career. Various seating arrangements were provided for 4, 6 or 8 people. Production ended in 1968 after thousands had been built for the West German army, the armies of NATO and civilian users. French forces in Germany still use some of them. They have permanent four-wheel drive and although often criticised, the all-independent suspension gives a good cross-country performance, especially in sand.*

The Ami with a Standard Vanguard engine was designed by Freighter Limited of Melbourne in 1954. It raised high hopes but never went into production.

The Gipsy, Austin's reply to the Land Rover came out in 1957 and production continued until 1968 but it was never a great success. Its most interesting feature was the Flexitor independent suspension by rubber, something between the Flector and the Silent-Bloc, but it was never satisfactory and from June 1965 steel springs were substituted. The Gipsy above with the short 90 in. wheelbase is a Series IV G4 M 10 with Austin 2,200 c.c. engine, petrol or diesel.

Since it was redesigned in the Sixties, the Toyota Land Cruiser has been very successful in various versions. It is built under license in Brazil as the Bandeirante. It is one of the most widely used cross-country vehicles in the world, mainly in the civilian sector. Oddly, the Japanese army does not use Toyotas, but Jeeps built under license by Mitsubishi. The Land Cruiser at the top of the following page was used by the photographic agency Sygma in the 1980 Paris-Dakar rally.

◀ *At the end of the fifties Bernard Cournil built the first of the "tractors" which bear his name. Strongly influenced by the Jeep it used some Jeep components for a long time, but was called a tractor because it was intended originally for agriculture. They are now made by Simi with petrol or diesel engines by Renault or Indenor and are built under licence in Portugal.*

3,922 of the Mighty-Mite were built by Amerian Motors for the US Marines between January 1960 and January 1969. The original engine was replaced by an American Motors V4 in aluminium. Some Mighty-Mites were used during the early years of the Vietnam conflict.

Having built the P2 from 1955 to 1962 in the Karl Marx plant, Barkas produced an improved model, the P3, until 1968. It had front wings like the West German Munga and only two doors. Suspension was independent by torsion bars and the engine developed 75 b.h.p. at 3,750 r.p.m. An amphibian version similar to the VW Schwimmwagen was also built.

In 1951 the US Ordnance Tank Automotive Command gave Ford a contract for a feasibility study on a new-country vehicle. Called the M151, it was produced by Ford and then by American Motors. In 1970 the front end was modified and the rear suspension, which had caused a number of accidents. The vehicle then became the M151 A2 seen here in 1977 during a Libyan army review in Tripoli.

In 1966 SINPAR presented three prototypes for a recreation vehicle called "Plein Air" based on the Renault R4 with four-wheel drive. The bodywork was in sheet steel but was replaced in production by one closer in style to the Renault 4L.

Seen at the Berlin Wall in April 1980, the P601 A built at Zwickau by Trabant is one of the standard all-terrain vehicles of the East German army although it has only two-wheel drive. The engine delivers 26 b.h.p. at 4,200 r.p.m. and the car weighs only 645 kg. A civilian version has been introduced under the name of Tramp.

Successor to the GAZ 69, the UAZ 469 B was shown by the Oulianovsk plant in 1961 but production in large numbers only began ten years later. Seen here in Cuba, 1978, it is one of the heaviest vehicles of its type at 1,500 kg.

The Peking is practically China's standard motor vehicle and looks generally similar to the UAZ 469 B. There are several versions, not exclusively for off-road use, and it is the most common Chinese motor vehicle. Powered by a 2.5-litre engine, it has a rather rustic finish, including the motor, and spartan accommodation.

One of the smallest all-terrain vehicles, the Suzuki is widely used by civilian owners. It looks like a war-time Gipsy Rose Lee.

A prototype developed by Citroen during the seventies using the mechanical elements of the Dyane 6.

Volkswagen Iltis with an elaborate front bumper in the second Paris-Dakar, 1980.

(Below) Introduced in 1974, the Fiat Nuova Campagnola is here seen in the 1980 Paris-Dakar. It has a two-litre engine and all-independent suspension. Several Italian and Swiss coachbuilders build luxurious bodies for it.

Designed and produced jointly by Mercedes-Benz in Germany and Steyr in Austria, the G Series is built with two lengths of wheelbase and a variety of body styles and engines. Unlike the Peugeot P4 it has locking differentials at front and rear, power steering, fixed windscreen and more luxurious interior.

At the Geneva motor show in 1980 Saurer introduced a new range of military vehicles, among them the F 006 developed from a prototype designed by Swiss designer Monteverdi.

With a wheelbase of 237 cm. the Jeep CJ 7 comes between the CJ 5 (212 cm.) and the CJ 6 (263 m.) This 78 model is the Golden Eagle version. American Motors introduced a CJ 8 with still more roomy bodywork in 1981.

Like all models in the range, the CJ 5 Renegade can be had with a four-cylinder engine of 2,5 litres, a six of 4,2 litres or V8 of 5 or 5,9 litres. Transmission can be manual or automatic and the Quadra-Trac limited-slip differential is available.

The latest model, the Laredo, is the most sophisticated of all the Jeeps.

Jeep is no longer just the name for a type of vehicle, it has become the trade name of the all-terrain vehicles built by American Motors who have thus managed to defend the name of "Jeep" which many other constructors wished to adopt. This Cherokee Chief is therefore a Jeep!

In 1980 American Motors marketed a Jeep diesel with Japanese Isuzu engine developing 65 b.h.p. at 3,800 r.p.m. after long trials at Yuma, Arizona. It has a range of 580 km. or 20 % more than the petrol model and is only available with four-speed manual gearbox.

LAREDO
Jeep

Photographic Credits

J.M. BONIFACE : 108, 109, 205, 213, 219, 220, 226. R.E.M.E. : 153. J. ROBINSON : 182. Ambassade de GRÈCE : 162. S. CHRULEN : 236, 244, 245. D. BELCHER : 243. SERVICE DÉPARTEMENTAL D'INCENDIE DU MAINE ET LOIRE : 175. POCLAIN : 237. SINPAR : 235, 236, 266. CITROËN : 203. RENAULT : 201. CMIDOM : 104, 105, 106, 107, 110, 112, 113, 114, 115, 118, 139, 181, 182, 190, 197, 211, 212, 215, 218, 229, 230, 231. FORD : 30, 37, 79, 144. PEUGEOT : 199, 200. PANHARD : 7, 204. FIAT : 12, 15, 260. BERLIET : 9. VOLKSWAGEN : 168. SMITHSONIAN INSTITUTE : 24, 25, 28, 38. WILLYS : 43, 44, 45, 71. VERARO : 235. DROITS RÉSERVÉS : 9, 10, 11, 15, 16, 19, 20, 21, 23, 26, 27, 29, 31, 32, 33, 34, 39, 40, 41, 46, 50, 52, 53, 55, 59, 60, 73, 77, 78, 79, 87, 89, 92, 119, 125, 126, 129, 138, 141, 142, 148, 155, 159, 161, 163, 165, 166, 171, 172, 173, 180, 182, 183, 184, 188, 189, 190, 192, 193, 198, 199, 201, 202, 206, 207, 208, 209, 210, 214, 227, 228, 229, 232, 233, 234, 236, 240, 248, 249, 250, 252, 253, 258, 260, 262, 263, 265, 267. ARCHIVES ISRAÉLIENNES VIA P.H. MERCILLON : 171, 184, 195. MUSÉE DES BLINDÉS SAUMUR : 12, 183, 198. A.F.P. : 14, 85, 86, 87, 139, 143, 149, 152, 170, 177, 186, 187, 191, 193, 195, 238, 239, 240, 267. ECPA : 17, 19, 54, 55, 69, 70, 71, 80, 84, 96, 106, 110, 116, 117, 131, 132, 133, 135, 136, 137, 141, 144, 145, 146, 147, 148, 149, 155, 169, 180, 185, 188, 189, 192, 193, 194, 196, 197, 203, 210, 212, 213, 214, 215, 216, 217, 219, 220, 222, 223, 225, 262. USIS : 72, 73, 74, 84, 90, 93, 94, 134, 136, 137, 138, 139, 140, 145, 146, 149, 150, 151, 152, 153, 157, 158, 178, 179, 181, 184, 233. IMPERIAL WAR MUSEUM : 8, 35, 36, 39, 42, 50, 51, 52, 74, 76, 78, 79, 80, 81, 82, 83, 84, 88, 89, 90, 91, 92, 94, 95, 96, 130, 131, 136, 139, 154, 156, 157, 158, 160, 161, 164, 165, 168, 169, 170. VANDERVEEN : 8, 12, 13, 17, 58, 59, 61, 62, 63, 64, 65, 66, 67, 68, 69, 70, 127, 132, 156, 161, 176, 191, 245, 263, 264. AEROSPATIALE : 217. VAN INGELGOM : 189. GENDARMERIE NATIONALE : 194, 220, 221. JAMESTOWNS FIRE DEPT : 174. SYGMA : 265, 268. AMBASSADE D'AUTRICHE : 194. DAF : 14, 261. SFERLAZZO : 35, 97, 98, 99, 100, 101, 102, 103, 111, 116, 119, 123, 241, 242. SIPA-PRESSE : 127, 213, 266. DOUGLAS AIRCRAFT : 151. GAZETTE DES ARMES : 205, 221, 266. PHOTOTHÈQUE ESSO : 236. CARROSSERIE ARNAULT : 241. KRENN : 122, 246, 247, 248, 255, 256, 257. F. TAINTURIER : 250, 251, 264. MAHEU-LABROSSE : 253. BIRO : 254. SIDES : 255. BRITISH LEYLAND : 259, 261, 264. ALFA ROMEO : 259. BENGSTON : 261. PEKIN AUTOMOBILE WORKS : 267. MERCEDES : 268. SAURER : 269. AMERICAN MOTORS : 269, 270, 271. ASSOCIATION LA ROSALIE : 120. G. NIEZER : 121. UNITÉ D'INSTRUCTION DE LA SÉCURITÉ CIVILE N° 7 : 124. A. HORB : 126. D. TARD : 127. DE LAFOURCADE : 128.

Acknowledgements

Societe Poclain, Monsieur Gouble
General Cavarot commanding the French Army information Service; Captain Vermande; Colonel Amet, SIRPA Gendarmerie.
CMIDOM, Monsieur Civita, chief Warrant officer Monerville; Monsieur Maurice Bouleau, Fabien Sabatès, Jacques Borge, Nicolas Viasnoff, Perkins Engines France, Monsieur Desmier; military Attachés at Greek and Austrian Embassies.
Colonel Aubry and Lt. Tributsch of the Saumur Armoured Vehicle Museum.
Messieurs Krenn, Bengston, Van Ingelgom, David Belcher,
Monsieur Farvacq of Ford-France, Pages of the Agence France Presse, Madame Josette Chardon of the Agence Sygma, Curators of the R.E.M.E and Paratroop Museums, Smithsonian Institute Museum.
Esso Photographic Library and Technical Section of the Ministry of Interior. Messieurs Maheu, Niesser, de La Fourcade, Horb, Martineau, Daubrosse, Stephane Ferrard and the Gazette des Armes, Patrick Mercillon, P. Chamourat, Michel le Rouziz, Dumont, Merinon and former staff of La Maltournée.
The Tour de France Organisers, James Town Fire Department, Civil Defence Training Unit N°. 7.
Chef d'Escadron (E.R.) Jean Robinson and Lieutenant Lefevbre.
Monsieur Castex and the Compagnie Generale de Geophysique, Madame Tatsis, Monsieur Kieffer, Gougaud, Jean-Charles, Lecarreres.
French Army Cinematographic Service, Citroen and Renault Commercial Vehicles, Panhard, Metamet.
Monsieur Duthilleul, the Rosalie Association for study of the history of fire brigades and their equipment, Lieutenant-Colonel Gauthier, Director of the Maine-et-Loire Fire Departement, Monsieur Victor Bouffort.
All our thanks also go to Monsieur Boucher.
We wish to thank Daniel Tard for the research he has done into the history of Hotchkiss and which he has kindly made available to the authors.
The authors offer their special thanks to Bart H. Vanderveen, through whose efforts and research a large part of history of the Jeep has been rescued from oblivion. His friendly assistance which has never been sought in vain has contribued a great deal to the compilation of this volume. We are extremely grateful.
The chapter on the history of the development of the Jeep in the United States owes much to methodical work of Jean-Michel Boniface to whom we offer our special thanks.

Achevé d'imprimer sur les presses de Berger-Levrault à Nancy, le 9 septembre 1981.
779276-9-1981 — Dépôt légal : 3e trimestre 1981. Printed in France.